LIST OF COVER PHOTOGRAPHS

Lower Left: The Apollo 11 rocket, carrying astronauts Armstrong, Aldrin, and Collins to the first Moon landing, is shown lifting off the launch pad from Cape Kennedy on July 16, 1969. The rocket was 363 feet tall, weighed 6,400,000 pounds, and was designed to send 100,000 pounds toward the Moon at 24,300 miles per hour. (NASA Photo)

Upper Left: A photo of the Moon taken by Apollo 17 when leaving lunar orbit. The left half of the Moon in this picture is part of the near side visible from Earth, while the right half is part of the far side which is never visible from Earth. The dark areas are the seas, or maria. The Sea of Tranquility, where Apollo 11 landed, is on the far left. (NASA Photo)

Upper Right: Apollo 16 astronaut John Young is shown at the peak of one of his jumps as he salutes the American flag on the Ocean of Storms. The Lunar Roving Vehicle and Lunar Module are in the background. (NASA Photo)

Lower Right: The Apollo 16 Lunar Module ascent stage is shown lifting off the Moon on its return to the Command Module. The photo was taken by a color TV camera left on the Moon. (NASA Photo 72-HC-273)

MOONGATE:

Suppressed Findings of the U.S. Space Program

The NASA-Military Cover-Up

By

William L. Brian II

FUTURE SCIENCE RESEARCH PUBLISHING CO.
P.O. BOX 06392 PORTLAND, OREGON 97206-0020

Copyright © 1982 by William L. Brian II

First Edition 1982

ALL RIGHTS RESERVED

No part of this book may be reproduced in any form or by any electronic or mechanical means including information storage or retrieval systems without permission in writing from the publisher.

Library of Congress Catalog Card No.: 81-69211
ISBN: 0-941292-00-2 Soft cover

Published by Future Science Research Publishing Company

Printed in the United States of America.

DEDICATION

This book is dedicated to the following groups and individuals:

The astronauts and other space program personnel who knowingly or unknowingly sabotaged the cover-up.

George Adamski, Howard Menger, and others who were courageous enough to share their experiences with the public in the face of unwarranted ridicule.

Researchers such as Immanuel Velikovsky, Baron Karl von Reichenbach, Wilhelm Reich, and others who have battled the status quo of science attempting to bring their incredible discoveries to light.

Beth Brian who typed the manuscript and contributed extensively to the editing.

All other supporters and contributors who value the truth.

ACKNOWLEDGMENTS

Appreciative acknowledgment is given to those groups and individuals who enabled portions of the following material to appear in this book:

From *The Mathematics of Space Exploration* by Myrl H. Ahrendt: Copyright© 1965 by Holt, Rinehart and Winston, Publishers: Reprinted by permission of Holt, Rinehart and Winston, Publishers.

From *Footprints on the Moon* by the Writers and Editors of The Associated Press with Manuscript by John Barbour: Copyright© The Associated Press 1969: Reprinted by permission of The Associated Press.

From "Moon's 'Eerie Sight,' Apollo Chief Says," by Howard Benedict: From *The Indianapolis News,* July 19, 1969: Copyright© 1969 by The Associated Press: Reprinted by permission of The Associated Press.

From *The Hollow Earth* by Dr. Raymond Bernard: © 1969 by University Books, Inc.: Published by arrangement with Lyle Stuart: Reprinted by permission of University Books, Inc.

From *History of Rocketry & Space Travel* by Wernher von Braun and Frederick I. Ordway III: Copyright © 1975 Wernher von Braun and Frederick I. Ordway III: Reprinted by permission of Harper & Row, Publishers, Inc.

From *Rendezvous in Space* by Martin Caidin: Copyright© 1962 by Martin Caidin: Reprinted by permission of E. P. Dutton & Co., Inc.

From "Space Travel" in *Collier's Encyclopedia,* 1961 ed.,: Reprinted with permission from Collier's Encyclopedia© 1961 Crowell-Collier Publishing Company.

From *The History of Rocket Technology* by Eugene M. Emme, ed.: Copyright© 1964 by Eugene M. Emme. Published by Wayne State University Press.

From "Interplanetary Exploration" in *Encyclopaedia Britannica,* 14th ed.: © 1960 by Encyclopaedia Britannica, Inc.: Reprinted by permission of Encyclopaedia Britannica, Inc.

From "Space Exploration" in *Encyclopaedia Britannica,* 14th ed.: © 1973 by Encyclopaedia Britannica, Inc.: Reprinted by permission of Encyclopaedia Britannica, Inc.

From "The Climb Up Cone Crater" by Alice J. Hall: From *National Geographic,* July 1971: Copyright © 1971 National Geographic Society: Reprinted by permission of The National Geographic Society.

From *Man and Space—The Next Decade* by Ralph E. Lapp: Copyright© 1961 by Ralph E. Lapp: Reprinted by permission of Harper & Row, Publishers, Inc.

From *Suiting Up for Space* by Lloyd Mallan: Copyright© 1971 by Lloyd Mallan: Reprinted by permission of Harper & Row, Publishers, Inc.

From *The Phantom of the Poles* by William Reed: Copyright 1906 by William Reed and Walter S. Rockey: Reprinted by permission of Health Research, Publishers, Box 70, Mokelumne Hill, California 95245.

From "What Is It Like To Walk on the Moon?" by David R. Scott: From *National Geographic,* September 1973: Copyright © 1973 National Geographic Society: Reprinted by permission of The National Geographic Society.

From "The Moon—A Giant Leap for Mankind" *Time,* July 25, 1969: Reprinted by permission from TIME, The Weekly Newsmagazine; Copyright Time Inc. 1969.

From *U.S. on the Moon* copyrighted 1969 U.S. News & World Report, Inc., Washington, D.C. 20037.

From *Principles of Astronautics* by M. Vertregt: All Rights Reserved 1965: Reprinted by permission of Elsevier Scientific Publishing Company.

CONTENTS

LIST OF PHOTOGRAPHS 13
PREFACE 17
CHAPTER 1 THE NASA-MILITARY CONNECTION 19
CHAPTER 2 LUNAR GRAVITY THEORY PRIOR TO
            MOON PROBES 29
CHAPTER 3 THE NEUTRAL POINT DISCREPANCY ... 39
CHAPTER 4 DISCREPANCIES IN SPACECRAFT
            VELOCITIES, FLIGHT TIMES, AND
            FUEL REQUIREMENTS 51
CHAPTER 5 ASTRONAUT EXPERIENCES ON THE
            MOON 61
CHAPTER 6 MOON ATMOSPHERIC THEORY PRIOR
            TO THE SPACE PROGRAM 83
CHAPTER 7 INCREDIBLE FINDINGS CONCERNING
            THE LUNAR ATMOSPHERE 89
CHAPTER 8 MOON GEOLOGY AND THE EARTH-
            MOON SYSTEM 113
CHAPTER 9 SATELLITE PHOTOS AND BIZARRE
            EARTH FINDINGS 129
CHAPTER 10 EVIDENCE OF EXTRATERRESTRIAL
            INTERFERENCE IN THE
            SPACE PROGRAM 145
CHAPTER 11 CONTACTEES AND MOON FINDINGS .. 157
CHAPTER 12 MALDEK AND THE MOON 169
CHAPTER 13 GRAVITY, SEARL DISKS, AND
            LEVITATING BEAMS................... 177

CHAPTER 14	THE PRESENT AND FUTURE SPACE PROGRAM	189

APPENDICES 197

FOOTNOTES 221

BIBLIOGRAPHY 227

LIST OF PHOTOGRAPHS

Photo 1 Bean standing next to Surveyor 3 with the Lunar Module in the background on the Apollo 12 Mission (NASA Photo)

Photo 2 Young jumping up from the lunar surface on the Apollo 16 Mission (NASA Photo)

Photo 3 Bean carrying the ALSEP on the Apollo 12 Mission (NASA Photo 69-HC-1341)

Photo 4 Irwin with the Lunar Rover and Mt. Hadley in the background on the Apollo 15 Mission (NASA Photo 71-HC-1140)

Photo 5 Aldrin standing next to the deployed Solar Wind Composition Experiment on the Apollo 11 Mission (NASA Photo)

Photo 6 The lunar surface with the Lunar Module and tire tracks on the Apollo 14 Mission (NASA Photo 71-HC-277)

Photo 7 Scott standing on the slope of Hadley Delta with the Apennine Mountains in the background on the Apollo 15 Mission (NASA Photo 71-H-1425)

Photo 8 Duke collecting samples on the rim of a crater with the Rover in the background on the Apollo 16 Mission (NASA Photo)

Photo 9 A halo around Bean on the lunar surface on the Apollo 12 Mission (NASA Photo 69-HC-1347)

Photo 10 A frame taken from the movie sequence of Mitchell's descent down the Lunar Module ladder on the Apollo 14 Mission (NASA Photo)

Photo 11 A picture of the Apollo 10 Command Module taken by the Lunar Module with the Moon's horizon in the background (NASA Photo)

Photo 12 Scott standing next to the Rover on the edge of Hadley Rille on the Apollo 15 Mission (NASA Photo 71-H-1426)

Photo 13 A picture taken by the author showing eroded hills in Eastern Washington State

Photo 14 A view of the eastern edge of the Sea of Rains with Hadley Rille and the Apennine Mountains taken on the Apollo 15 Mission (NASA Photo)

Photo 15 Schmitt standing next to a split boulder with rounded hills in the background on the Apollo 17 Mission (NASA Photo)

Photo 16 The Moon's Alpine Valley photographed in 1967 by the Lunar Orbiter IV probe (NASA Photo 67-H-897)

Photo 17 An Applications Technology Satellite photo of the Earth taken 22,300 miles above Brazil in 1967 (NASA Photo 67-HC-723)

Photo 18 A reproduction of the color TV transmission of the Lunar Module lift-off from the lunar surface on the Apollo 17 Mission (NASA Photo 72-HC-903)

Photo 19   A reproduction of the color TV transmission of the Lunar Module lift-off from the lunar surface on the Apollo 16 Mission (NASA Photo 72-HC-274)

Photo 20   A reproduction of the color TV transmission of the Lunar Module lift-off from the lunar surface on the Apollo 16 Mission taken moments after Photo 19 (NASA Photo 72-HC-273)

PREFACE

Moongate is a non-fictional account derived from unclassified sources of information including official government publications, NASA photographs and movies, news articles, and numerous books by authorities in various fields. The idea for *Moongate* germinated from observations made by several individuals who detected discrepancies in space program activities and findings. An extensive search for evidence was made to verify that a cover-up occurred. The amount of evidence far exceeded initial expectations and produced startling conclusions about the real space program.

Many aspects of the space program are mathematically and conceptually analyzed to verify the cover-up from a scientific standpoint. *Moongate* is written for the layman and scientist alike, with mathematical calculations included in the appendices. It contains many color photographs and footnoted references.

Although absolute certainty cannot be given about the details of the cover-up, the evidence demonstrates that either the contentions are true, or portions of the Apollo Moon landings were staged. It seems probable that the landings really occurred; however, the true circumstances surrounding the Apollo missions and related discoveries were carefully suppressed from the public.

William L. Brian II

CHAPTER 1

THE NASA-MILITARY CONNECTION

The NASA space program to put men on the Moon was supposedly an unclassified, civilian effort. However, evidence will be provided throughout this book that the military had almost complete control over it and that many of the NASA findings were withheld from the public.

The history of rocket development and early satellite efforts will now be reviewed to show the military's involvement in the space program and events leading to the creation of NASA. Germany developed the V-2 rocket and used it with limited success against England toward the end of World War II. The United States acquired the majority of German rocket scientists and engineers, including Wernher von Braun, and most of Germany's rocket hardware. Russia also managed to obtain German rocket personnel and pursued the technological development of rockets with great enthusiasm and success. The postwar years ushered in the "cold war" and the threat of nuclear war by the superpowers. The rocket was then developed to its high level of sophistication to carry nuclear bombs to strategic targets thousands of miles away. Typically, the threat of annihilation seems to be the greatest stimulus to expend large dollar amounts on research and development, and the rocket was no exception.

The confiscated German rocket information included plans for Earth satellites and multi-staged rockets which would have eventually been fired at North America from Europe if the conflict continued. This caused speculation on the future of warfare and provided incentive for development of an advanced rocket technology. However, the United States did not seriously engage in rocket research until the hydrogen bomb was developed in 1952 and until the discovery was made in 1953 that the Russians were well along in developing rocket missiles. Supposedly, the latter discovery caused the U.S. Government to reverse its previous position on the importance of the ICBM (intercontinental ballistic missile). In 1953, the Russians tested their own H-bomb, and in 1954 a policy was approved by the National Security Council to prioritize ICBM development. By that time, the Russians had already designed an oversized carrier rocket to deliver the heavy A-bomb. It had excess capacity for the lighter H-bomb and Soviet scientists supposedly suggested that it could be used to launch satellites into orbit.

The U.S. had anticipated the Soviet idea of artificial satellites in an Army Air Force satellite study conducted in 1946 called Project Rand. Project Rand was staff by a consultant group managed by Douglas Aircraft Company. Their report was entitled *Preliminary Design of an Experimental World-Circling Spaceship.* Included in the report were a consideration of the technical feasibility, the political and psychological effects of it as a weapon, its use as a surveillance and communication device, and a demonstration of U.S. technological superiority. Since the Germans had already thought of it, the Rand Report and the Russian plan cannot be considered original. In actuality, the U.S. and Russia merely extended German plans, all in the interest of military supremacy.

In 1946, the Navy and Army Air Force conducted independent missile studies. A second Project Rand study published in 1947 gave more detailed specifications for a three-stage rocket to put satellites into orbit. This study also indicated some of the support technology areas which needed additional development. Included were guidance and flight control, orbital attitude control, ground-space communications, and auxiliary power sources. Essentially, the primary need was for miniature computers and solar energy power supplies which were not available then. In 1947, the Secretary of Defense was added to the cabinet and the Research and Development Board (RDB) was placed under the Department of Defense. The RDB postponed the decision of which military branch would develop and deploy long range missiles. Through 1948, the Navy continued their satellite research activity while the Air Force discontinued theirs.

The first public announcement of the satellite program was included in the Report of the Secretary of Defense in 1948. In *The History of Rocket Technology,* written in 1964, Cargill Hall states:

Forrestal's early veiled announcement of the U.S. satellite program caused a great deal of consternation among those working on the satellite project in the United States in 1949 and who were trying to preserve its aspect of secrecy. . . . After this occurrence, *no official reference* in open literature was made to an American satellite effort until November 1954, when the Department of Defense, in a terse two-sentence comment, reported that studies continued to be made in the Earth Satellite Vehicle Program. The statement, approved by Secretary of Defense Charles Wilson, was issued after an earlier press conference statement by the Secretary that he was unaware of an American Satellite program.[1]

The Department of Defense budgeted over $1 billion for missile programs in 1952 which nearly equaled the sum of expenditures over the previous five years. This was spent primarily on short range, surface-to-surface, and air defense missiles. However, the Army was working on a larger missile under the direction of Dr. Wernher von Braun who was then the technical director of the Army Ordnance Guided Missiles Development Group. In 1951, he proposed the building of a 7,000-ton satellite rocket. Supposedly the military was not interested in satellites, so U.S. scientists tried to get civilian agencies to promote a satellite program. Soviet participation in scientific conferences in 1954-1955 gave U.S. scientists the impression that Russia was involved in a vigorous space program. In 1955, the White House announced that plans had been approved for an Earth-circling satellite for scientific observations. It seems that most of the public was never aware that Secretary of Defense Forrestal announced the U.S. satellite program in 1948.

Sputnik I was fired on October 4, 1957 and Congressional investigations into the U.S. missile and satellite program followed. Typically, the "surprise" efforts of Sputnik I were forecast by Project Rand in 1946, yet the U.S. Congress and public were not involved until 1957. This is representative of the secrecy of the military in their operations and of the delay in information reaching the public.

A lot of publicity was drummed up for Project Vanguard which was to have its first launching in 1957. Unfortunately, Vanguard I blew up on the launch pad at Cape Canaveral on December 6, 1957. However, in Huntsville, Alabama on January 31, 1958, Wernher von

Braun and his rocket team successfully sent Explorer I into orbit using the Jupiter C rocket.

NASA was created on October 1, 1958 to coordinate national space activities as a result of President Eisenhower's April 2 message to Congress. One of the urgent reasons given to expand the space program was to take full advantage of the military potential of space. NASA was to administer the "civilian" space science and exploration program.

The lengthy Congressional investigations into the "missile-gap" served to convince Congress of the importance in keeping up with the Russians. However, it seems that the aspect of enhancement of national prestige was more appealing to politicians than the military potential of space. In addition, the background of politicians does not generally lend itself to an appreciation of scientific research and space exploration. Therefore, the space race aspect of the space program was emphasized because it was something to which the politicians could relate. It also served as a means for building up the momentum which would be needed to finance the expensive space program. The "civilian" space program was really set up to take emphasis off the military aspect of the project and to develop technology for military applications at the same time. By maintaining a quasi-civilian organization like NASA, public financial support could be gained and the work could be accomplished more efficiently.

Concerning the Defense Department's involvement, Ralph Lapp stated the following in *Man and Space— The Next Decade,* written in 1961:

> The Defense Department did have a legitimate stake in the satellite field. There was a military requirement for orbital devices that could perform

communications and reconnaissance missions. . . . the development of the Atlas ICBM with its 360,000-pound thrust gave the Defense Department the potential of boosting heavier payloads into space. . . . Naturally, the reconnaissance or "spy" satellite program was highly classified. This fact, together with the military secrecy attached to ICBM rockets which would be used for civilian applications, added complexity to the emergency U.S. space program.

A purely peacetime civilian space activity could be conducted most effectively and most efficiently if all details of the work could be kept unclassified. . . . Unfortunately, it seemed, even space science was to have two antipodal aspects.

In one respect, it seemed that the new U.S. space agency would have a clear field, unfettered by military bonds. The Pentagon saw no need for spaceships or for the huge rocket engines needed to power them. This was to prove a lucky break for the civilian space program, even though it later developed that there could be military applications for very high-thrust rockets.[2]

It is important to note that the military continued with its own efforts in satellites and missiles throughout the peak era of NASA and into the present. Although NASA is only a fraction of its former size, the military is as strong as ever. It continues to utilize the "civilian" research and development information and hardware developed through the space program for its own secret projects.

The funding of top secret military projects can be easily camouflaged by other unclassified projects. Components and parts needed for secret projects can be ordered one at a time from different manufacturers and charged to a dummy or decoy project. The parts can be assembled in secret and the manufacturers never know what the end product is. An expensive decoy project can be used to generate the funds and to develop the technology for highly sophisticated secret projects at the same time. The NASA space program to put men on the Moon provided the military with such a decoy project.

Military secrecy related to research and development of weapons has existed for a long time. The government's rationale for this policy of total secrecy has been to maintain superiority over the enemy. A side effect is that the public is automatically kept in complete ignorance of what the military is doing, so a great deal of horrifying research can be carried out without public interference. Secrecy is deemed necessary for survival and the public is always kept many years behind the latest research findings and technological developments. When secret information is finally released, the government is always quick to explain that the defense imperative required total secrecy for our protection.

To silence top secret research participants, secrecy acts can be used. If the secrecy acts are breached, the violator can be judged to be insane and committed to a mental institution, be sent to prison, be paid to keep quiet, or perhaps meet with an unfortunate accident if all other persuasions fail. Since the author has signed no secrecy agreements with anyone, he is under no obligations. If he merely points out what already exists in the available, unclassified literature, the government can only accuse him of idle speculation or remain silent. If a smear campaign ensues, it will not come as a surprise. The government has had a lot of experience in this area and has practically unlimited resources and government agencies to carry it out.

In the remainder of this book, it will be made painfully apparent to the reader that the military was in constant control of the entire NASA space program. It will be shown that much of the NASA-related information was highly classified. The public was only given enough information to convince most of them that men had landed on the Moon. The majority of details and discoveries regarding the project were carefully suppressed.

It is interesting to analyze the people who want to develop the military potential of space. Some of these undoubtedly have the "mega-death" type of mentality. They seek better and better methods to destroy life, that is, more efficient methods, always developing strategies and weapons which are superior to the enemy's strategies and weapons. They would exploit the military potential of the Moon as soon as the technology were available. Eventually, other planets would become military outposts. Finally, artificial battle stations, like those seen in the movie *Star Wars,* would be constructed. However, if advanced intelligences exist in the universe which are superior to Earth men, the militarists will eventually meet their match in outer space. Evidence will be presented later that this may have already happened.

In closing this chapter, it has been suggested that NASA and the military suppressed new discoveries and findings of the space program. If the cover-up had been totally successful, this book could not have been written. When thousands of people are involved in a project the size of the space program for many years, airtight security would be nearly impossible. In addition, a number of individuals are basically honest and want to tell the truth, in spite of pressures to remain silent.

The NASA-Military Connection

Chapter 2 delves into a fundamental law defined by Isaac Newton in 1666 concerning gravity. It will be shown that this law was found to be incorrect when applied to planetary bodies. Consequently, the first attempts to explore the Moon with space probes produced unexpected results.

CHAPTER 2

LUNAR GRAVITY THEORY PRIOR TO MOON PROBES

According to conventional science, the Moon has only one-sixth of the Earth's surface gravity. Sir Isaac Newton formulated the Law of Universal Gravitation in 1666 which led to this conclusion. This famous law states that the gravitational pull of one body on another body depends on the product of the masses of the two bodies. Therefore, a planet such as the Earth will attract another object with a certain force. The law also states that the gravitational force decreases as the distance away from a planet increases. The further out in space an object is from the Earth or Moon, the less pull or tug is exerted on it.

Newton discovered that gravity decreases away from the surface of the Earth in the same way light intensity diminishes away from a light source. Only one-fourth as much light illuminates a given surface area 200 feet from a light source as compared to 100 feet. Similarly, at 300 feet only one-ninth as much is received as compared to 100 feet. This rapid drop-off follows the inverse-square law. For the following discussion, see Figure 1.

Near the Earth's surface, objects fall with an acceleration of 32.2 feet per second every second. Therefore, as each second goes by, an object will gain in speed by 32.2 feet per second. It will continue to accelerate until it reaches a constant speed because of air resistance.

Figure 1

DECREASE IN GRAVITY WITH DISTANCE AWAY FROM THE SURFACE OF THE EARTH

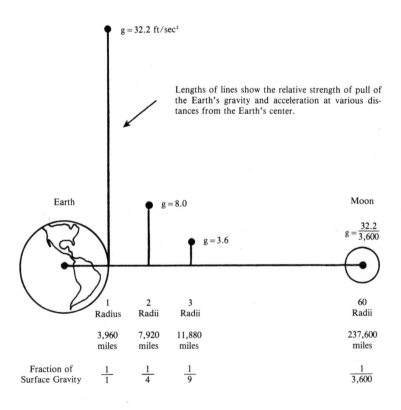

Lunar Gravity Theory Prior to Moon Probes

In moving away from the Earth's surface, an observer will find that at an altitude of 3,960 miles, or two radii from the Earth's center, the gravitational pull has been reduced by a factor of four, just as in the light example. At this altitude, bodies have only one-fourth of their surface weight; therefore, a 200-pound man would weigh only 50 pounds. Furthermore, he would be accelerated at one-fourth the surface rate, or 8 feet per second every second, and it would take him twice as long to fall a given distance starting from rest.

Moving out to a distance of three Earth radii from the center, or 7,920 miles from the surface, the force of gravity decreases to one-ninth of the surface value. The 200-pound man would now weigh only 22 pounds, and would only be falling 3.6 feet per second faster after every second of fall.

At the distance of the Moon, the force exerted by the Earth is only 1/3,600 the value on the Earth's surface. Therefore, the Moon should only be falling toward the Earth 32.2/3,600 feet per second faster every second. It would soon crash into the Earth if it were not slowly rotating about the Earth every 27 days or so. This orbit, or rotation, is what keeps it from falling. Man-made satellites orbit the Earth in the same way the Moon does. However, since they are usually much closer to the Earth than the Moon, the Earth's pull is significantly stronger on them and they must travel much faster to continue orbiting. The Moon moves around the Earth at a speed of about 2,300 miles per hour, while a satellite 100 miles above the Earth must travel nearly 17,500 miles per hour.

Newton's analysis of gravitation was devised from the observation of the orbiting Moon and objects falling on the Earth. Until similar experiments are performed on planetary bodies like the Moon, the exact values of their surface gravities cannot be determined. Newton could not even determine the Moon's mass to predict the Moon's force of attraction on other objects. Its mass was later calculated to be about one-eighty-second (1/82) of the Earth's by observing how much the Earth moves around the common center of rotation of the Earth and Moon. In turn, the Moon's derived mass and the Earth's predicted mass were used to calculate the Moon's surface gravity which came out to be one-sixth of Earth's. Since the Moon is a much smaller body than Earth, it did not seem unreasonable to scientists that it should have a correspondingly smaller surface gravity.

The point where a space vehicle enters the predominant attractive zone of the Moon's gravity is called the neutral point. It is the region in space where the Earth's force of attraction equals the Moon's force of attraction. Since the Moon is smaller and supposedly has a smaller surface gravity, the neutral point should be quite close to the Moon. In fact, if it is assumed that the Moon has one-sixth of the Earth's surface gravity, the neutral point is calculated to be about nine-tenths of the distance between the Earth and the Moon. The average distance to the Moon is about 239,000 miles, hence this places the neutral point approximately 23,900 miles from the Moon's center. Figure 2 depicts the neutral point. To show the reader that this neutral point distance has been predicted and calculated time and time again by astronautical scientists and engineers for many years, a series of references will be given.

In the book *Principles of Astronautics,* written in 1965, a fellow of the British Interplanetary Society, M. Vertregt, gave the calculated neutral point value as follows:

Figure 2

THE CONVENTIONAL POSITION OF THE NEUTRAL POINT BETWEEN THE EARTH AND THE MOON

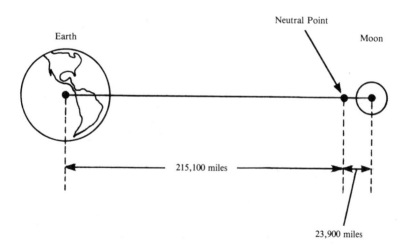

At a distance of 346000 km (215000 miles) from the Earth and 38000 km (23600 miles) from the Moon, at what is called the "neutral point" N, the attraction exerted by the Earth is equal to that exerted by the Moon.[1]

In *Exploration of the Moon* written in 1966 by an astronomer, Franklyn M. Branley, the neutral point is listed as 20,000 miles from the Moon or 220,000 miles from the Earth.[2]

In *U.S. on the Moon* written by *U.S. News & World Report* in 1969, the neutral point distance is again displayed as 22,000 miles from the surface of the Moon.[3]

In *The Mathematics of Space Exploration* written in 1965, Myrl H. Ahrendt calculated the neutral point using Newton's Gravitation Law and the Moon's mass expressed as 1/83 of the Earth's mass. His conclusion, assuming a Moon to Earth distance of 239,000 miles was:

> ... the neutral point is about 23,900 miles from the Moon, at a point almost precisely nine-tenths of the distance to the Moon.[4]

Still another derivation of 23,800 miles was made by John A. Eisele in 1967 in *Astrodynamics, Rockets, Satellites, and Space Travel,* assuming the Moon's mean distance of 238,857 miles.[5] He assumed a ratio of Earth to Moon masses of 81.56.

Within the topic "Space Travel" in the 1961 edition of *Collier's Encyclopedia,* the following is stated:

> ... there must come a point where the two pulls are equally strong and where they balance each other. This point lies about 23,500 miles from the surface of the moon.[6]

The 1960 printing of the *Encyclopaedia Britannica*

states the following within the topic "Interplanetary Exploration":

... the so-called "neutral point" between Earth and Moon. This is a fictitious station on the Earth-Moon axis (about 19 Moon radii from the Moon), beyond which the Moon's gravitational pull is stronger than the Earth's.[7]

Nineteen Moon radii equals 20,520 miles from the Moon.

It is probably evident to the reader that minor differences exist between these values. This is due to slightly different assumptions of the Earth to Moon distance and ratio of Earth to Moon masses. An analysis of how much this neutral point distance should vary, depending on the Earth to Moon distance, gives the results shown below. This assumes that the distances are measured from the Earth's center to the Moon's center.

Total Distance Center to Center (miles)	Earth to Neutral Point (miles)	Moon to Neutral Point (miles)
252,710	227,517	25,193
238,885	215,070	23,815
221,463	199,385	22,078

In any case, the range of neutral point distances to the Moon's center will be between 22,078 and 25,193 miles with the assumption that the Moon has one-sixth of Earth's surface gravity. Since so many people and organizations have stated the neutral point distance within a close range, there seems to be no question as to where it is. To satisfy the technically minded reader, the complete derivation of the neutral point distances shown above is presented in Appendix A.

The reader is reminded that the above neutral point distances were based upon Newton's Law of Universal

Gravitation. In addition, most of the writers who referenced them were probably unaware of space program findings regarding the neutral point's real location. As mentioned previously, only by observing falling or orbiting bodies in the Moon's vicinity could the actual neutral point distance, hence the Moon's true gravity, be determined. This information could have been available to NASA or the Russians as early as 1959 from the first Moon probes. If Moon probes successfully orbited and landed prior to 1969, the actual neutral point distance should have been available to the public. If it were published, there would be an exact method of determining the Moon's surface gravity.

The ramifications of finding a neutral point distance greater than the 20,000 to 25,000-mile distances referenced above will now be considered. It was shown how gravity falls off with distance away from the Earth's surface. The Moon displays the same behavior as the Earth. Therefore, the Moon's gravitational pull 1,080 miles above its surface (two radii from the Moon's center) would be one-fourth the surface value. Similarly, at three Moon radii, 2,160 miles above the surface, the force would be one-ninth the surface value. With this concept in mind, the reader can see that if the real neutral point distance from the Moon's surface were considerably greater than 25,000 miles, the Moon's surface gravity would have to be much greater than the predicted value of one-sixth of the Earth's surface gravity. This would imply that Newton's Law of Gravitation does not hold for bodies of planetary size. It would also imply that NASA and the military have been suppressing information on the true nature of Moon gravity.

Clearly, this neutral point distance had to have been exactly determined if astronauts were to land safely on the Moon. This could only be determined experimentally and a history of this discovery will be presented in the next chapter.

CHAPTER 3

THE NEUTRAL POINT DISCREPANCY

A lunar probe or spacecraft launched from the Earth will continuously lose velocity until it reaches the neutral point due to the Earth's gravitational pull. However, after it passes the neutral point, the Moon's pull becomes stronger and it begins to accelerate, increasing in velocity. It must have the proper trajectory to assume a lunar orbit or to score a direct hit.

The need for an accurate measurement of the Moon's gravity, hence the precise neutral point distance, was pointed out by Hugh Odishaw, Executive Director of the United States National Committee for the IGY (International Geophysical Year). He presented a report in 1958 to all member nations of the IGY entitled "Basic Objectives of a Continuing Program of Scientific Research in Outer Space."[1] In it he indicated that estimates of the Moon's mass at that time were based on observations of the motions of asteroids and the Earth's polar axis. The uncertainty attributed to the Moon's mass was given as 0.3 percent which was great enough to affect lunar rocket trajectories. Accordingly, Odishaw indicated the desirability of determining the Moon's mass more precisely in early Moon experiments. This could be accomplished by tracking the rocket as it approached the Moon and deriving the Moon's pull at each point of the trajectory, hence the surface gravity.

By now, the reader probably realizes how much difficulty NASA and the Russians would have had in sending successful Moon probes, even if they knew the exact position of the neutral point. If the neutral point, hence the Moon's gravitational pull, deviated considerably from the predicted value derived from Newton's Law of Universal Gravitation, a series of failures would be expected in attempts to send successful lunar probes. It is also reasonable to conclude that a discovery of a significant difference in the expected Moon gravity would require many more years of reprogramming, rocket design, lunar probe design, and so on. The time required for people to readjust their thinking patterns would also be significant, especially after nearly 300 years of education and training in the gravitational concepts of Isaac Newton. In the style of the Department of Defense, it should also be expected that suppression of the new findings would occur. Keeping these ideas in mind, along with the conventional idea of the position of the neutral point from the Moon, the history of lunar probes will be reviewed.

The Moon was chosen as the first target for exploration because it is the closest celestial body to the Earth. Russia was the first nation to send a successful lunar probe, called Luna 1, on January 2, 1959. It flew within 4,660 miles of the surface and broadcast information back to Earth after traveling into space. The U.S. had made three unsuccessful attempts with Pioneers 1, 2, and 3 in 1958 before achieving a fly-by 37,300 miles from the surface several months after Luna 1.

Luna 2 was launched on September 12, 1959 and became the first lunar probe to hit the Moon, sending back signals before impact. Luna 3 was launched October 4, 1959 and circled behind the Moon, approaching within 4,372 miles. It sent back pictures of the far side. Significantly, the Russian program for exploration of the Moon came to a stop for four years following the

The Neutral Point Discrepancy

Luna 3 lunar probe! All of the Luna shots were tracked with radar to collect trajectory and gravitational data. As previously mentioned, the trajectory of an object in the Moon's vicinity enables the surface gravity to be calculated which in turn enables the neutral point to be calculated. If the findings deviated from the expected ones, it would probably require years to reassess and reengineer future Moon probes. A soft-landing would require a much larger launch vehicle and a great deal more fuel if the gravity were a lot higher than expected.

Russia's secrecy concerning its space program is well-known. Therefore, the U.S. may not have benefited from information obtained by Russian Moon probes. According to Ralph Lapp in *Man and Space—The Next Decade:*

. . . the Soviets clamped tight secrecy over their rockets, never once releasing a photograph of a launching. Moreover, the Russian scientists were slow in making their data available to the scientific community.[2]

In addition, the U.S. Pioneer 4 fly-by at 37,300 miles may not have been close enough to the Moon to enable NASA engineers to determine the true nature of lunar gravity. At any rate, subsequent Ranger missions indicated that the U.S. was having many problems in achieving successful Moon shots.

The first Rangers carried seismometers in spherical containers designed to withstand the impact of landings. Unfortunately, Ranger 3, launched on January 26, 1962, missed its target completely and went into a solar orbit. Ranger 4 hit the Moon on April 23, but did not send back any useful information. Ranger 5 was launched on October 18 and missed the Moon by 450 miles; however, it was tracked for over eight hours. Further launches were put off until 1964 and the entire program was reorganized.

It is significant that all Ranger missions after number 5 were designed only to take pictures because of the difficulty in achieving a semihard landing with the seismometer package. The seismometer was encased in a 30-inch balsa wood ball which was to be slowed up sufficiently by retro-rockets to hit the surface at 150 miles per hour and still survive. It was designed to be able to impact granite at 200 miles per hour and continue to operate. If the Moon had only one-sixth of Earth's surface gravity, then perhaps the seismometer packages would have survived. However, if lunar gravity were much more than expected, a successful landing without big enough retro-rockets for braking would be impossible. Evidently, Ranger scientists anticipated that the weak one-sixth gravity would keep the velocity of impact down to a low enough level. Since they eliminated the package from further missions and delayed these missions for almost a year-and-a-half, perhaps they learned something new about the Moon's gravity.

After Russia's four years of silence, Luna 4 was launched on April 2, 1963. It flew within 5,300 miles of the Moon. The purpose of this probe was never revealed except for a brief announcement that:

> ... experiments and measurements which were conducted ... are completed. Radio communication with the spacecraft will continue for a few more days.[3]

It is probable that the need for detailed gravity data was behind the mission. Soft-landings could not be successful without this information.

The U.S. launched Ranger 6 on January 30, 1964 and the electrical system was allegedly burned out when the cameras were accidentally turned on during the flight,

The Neutral Point Discrepancy 43

hence no pictures were sent. After supposedly redesigning the system to eliminate this danger, Ranger 7 was launched on July 28. It was successful and sent back thousands of pictures. Ranger 8 was launched on February 17, 1965 and Ranger 9 was launched on March 21, 1965. Both were successful and some of the Ranger 9 pictures were broadcast on television.

The Russians attempted a soft-landing with Luna 5 on May 9, 1964, but it crashed at full speed. Luna 6 was launched on June 8 but missed the Moon, while Luna 7 crashed because the retro-rockets supposedly fired too soon. Luna 8 was sent up on December 3 and also crashed. Luna 9 landed successfully on the Moon on February 3, 1966.

The U.S. soft-landing program was called Surveyor and began in 1960. In 1962 a decision was made to trim the weight of Surveyor by more than 300 pounds, with many experiments abandoned. The reason given was problems with the proposed Atlas Centaur second stage. Surveyor's scheduled 1963 launch date passed and it was not even close to being ready. The project costs were running ten times the original estimates and "troubles" forced one delay after another. A Congressional inquiry was made and the House Committee on Science and Astronautics found fault with the management practices of the Jet Propulsion Laboratory (JPL), NASA, and the prime contractor, Hughes Aircraft. In *We Reach the Moon,* John Noble Wilford gave an account of the Surveyor difficulties.[4] It seems JPL officials conceded that they initially underestimated the difficulty of the project. One official admitted that the project was not given enough support in the earlier days and that they were overconfident in their ability to do things. It is probably more than coincidental that the Ranger 5 failure on October 18, 1962 resulted in the abandonment of the seismometer package and a significant delay in future Ranger missions -due to the difficulty in a semihard landing. The Surveyor Program was delayed for 28 months from its schedule and Surveyor 1 did not soft-land on the Moon until June 2, 1966. Photo 1 shows Apollo 12 astronaut Alan Bean standing next to Surveyor 3 which landed April 20, 1967 inside a crater in Oceanus Procellarum. The Apollo 12 Lunar Module is in the background on the rim of the crater.

The U.S. effort to orbit the Moon using lunar probes began August 17, 1958 with Atlas Able 1. It missed the Moon as did the next two attempts. A decision was then made to build a larger spacecraft and to use the Atlas Agena D as the carrier. It appears that a larger rocket was necessary to carry a larger payload which may have consisted of fuel used in braking the proposed orbiter. This would be necessary to reduce the velocity of the satellite so that it could achieve an orbit. Again, it seems more than coincidental that the project to orbit the Moon, which began in 1958, was postponed until 1964 when the Boeing Company began work on the Lunar Orbiter Project.

The Russians managed to place Luna 10 into orbit around the Moon on April 3, 1966 after having successfully soft-landed with Luna 9 on February 3, 1966. It appears that substantial retro-rocket braking was required for orbit insertion as well as soft-landing. At any rate, both were accomplished a short time apart. U.S. Lunar Orbiter 1 successfully went into lunar orbit on August 14, 1966. Lunar Orbiter 5 was sent crashing into the Moon on January 31, 1968 after a successful mission. These missions photographed over 99 percent of the Moon and led to the discovery of lunar mascons, or

The Neutral Point Discrepancy

increases in the Moon's surface gravity in certain areas. These mascons will be discussed in more detail later.

The above analysis of lunar probes indicates that the U.S. and Russia probably had a clear picture of the nature of lunar gravity as early as 1959. However, it is a certainty that both countries learned how to work with lunar gravity and make soft-landings by 1966. This date is important in light of information on lunar gravity to be presented next.

The reader has been kept in suspense concerning suggestions that Moon gravity might deviate from the predicted value of one-sixth of Earth's. This was necessary to provide background information needed to make a proper evaluation. The analysis will now focus on the position of the neutral point, as given to the public by various writers and organizations, subsequent to lunar probes. Ultimately, the source of the information is probably NASA. In reference to Apollo 11, *Time* magazine gave the following neutral point information in the July 25, 1969 issue:

> At a point 43,495 miles from the moon, lunar gravity exerted a force equal to the gravity of the earth, then some 200,000 miles distant.[5]

The reader might be surprised concerning this statement since the neutral point distances presented in Chapter 2 were all 20,000 to 25,000 miles from the Moon. It might seem that *Time* has made an error; therefore, other sources will be pursued to verify this figure.

In the 1969 edition of *History of Rocketry & Space Travel* by Wernher von Braun and Frederick I. Ordway III, the following statement is made concerning Apollo 11:

> The approach to the Moon was so precise that the mid-course correction scheduled for 8:26 a.m. (EDT)

on the 19th was canceled. At a distance of 43,495 miles from the Moon, Apollo 11 passed the so-called "neutral" point, beyond which the Lunar gravitational field dominated that of Earth. Consequently, the spacecraft, which had been gradually losing speed on its long coast away from Earth, now began to accelerate.[6]

Note that the precision of the flight was so great that the mid-course correction was not needed. In addition, the neutral point distance is given to the nearest mile and agrees exactly with the value given previously by *Time* magazine.

Another reputable source is the *Encyclopaedia Britannica*. This organization generally publishes information which is accepted by orthodox scientists. Therefore, their claim for the neutral point distance should be in close agreement with Wernher von Braun. In reference to Apollo 11, the *Britannica* stated the following in the 1973 printing within the topic "Space Exploration":

Consideration of the actual dynamics of the Apollo trajectory will review the statements made above. The Apollo 11 spacecraft had been in Earth orbit at 118.5 mi. altitude, traveling at 17,427 mph. By firing the rocket motor at the exact moment when the spacecraft was precisely aligned along the proper trajectory, the velocity was increased to 24,200 mph. Because the Earth's gravitational pull continued to act upon the spacecraft during its two and three-quarters day (64 hr.) journey toward the Moon, the spacecraft velocity, with respect to the Earth, dwindled to 2,040 mph at a distance of 39,000 mi. from the Moon. At this point lunar gravitational attraction became greater than the Earth's and the spacecraft

commenced accelerating as it swung toward and around the far side of the Moon, reaching a speed of 5,225 mph. By firing the spacecraft rocket propulsion system the velocity was reduced to 3,680 mph and the spacecraft entered an elliptical orbit about the Moon.[7]
Here the distance is 39,000 miles which is still close to the values given by *Time* magazine and von Braun. The reader may recall that in Chapter 2 reference was made to the 1960 printing of the *Encyclopaedia Britannica* which listed the neutral point distance as 19 Moon radii, or 20,520 miles, from the Moon. In this case, the distance discrepancy is between different printings of the same source.

In *We Reach the Moon,* Wilford indicated that the spacecraft entered the lunar sphere of gravitational influence about 38,900 miles from the Moon.[8]

In *Footprints on the Moon* written in 1969 by the Writers and Editors of the Associated Press, the neutral point is described as follows:

Friday, Day Three of the mission, found Apollo 11 at the apex of that long gravitational hill between earth and the moon. At 1:12 p.m. EDT, the nose-to-nose spaceships passed the milestone where the moon's gravity becomes the more important influence. The astronauts were 214,000 miles from earth, only 38,000 miles from their rendezvous with the moon, leading their target like a hunter leads a duck.[9]

The reader may already recognize the inconsistencies between the quoted figures which vary between 38,000 and 43,495 miles. Many different values are given with varying degrees of precision, yet they still lie within a range which is radically different from pre-Apollo calculations. There is no way to get around the discrepancy between the conventional, pre-Apollo distances of 20,000 to 25,000 miles, and the post-Apollo range of 38,000 to 43,495 miles. Even though the Earth to Moon distance varies between 221,463 and 252,710 miles, and spacecraft do not travel on a straight line between the Earth and Moon, this still does not explain the neutral point distance discrepancy. The logical conclusion is that the latest neutral point information reached the general public at about the time of the first Apollo lunar landing in 1969, even though it was determined as far back as 1959 from early lunar probes. Clearly, this discrepancy has not been pointed out to the public until now. To this day, the status quo of science and government alludes to the one-sixth gravity of the lunar surface, representative of a neutral point less than 25,193 miles from the Moon. Therefore, the neutral point discrepancy and its implications must be investigated.

The Moon's surface gravity was calculated with the new figures presented above using the standard inverse-square law technique. Since the radii of the Earth and Moon, the neutral point distance, and the Earth's surface gravity are known, the Moon's surface gravity is easily determined. The technique does not require a knowledge of the Moon's mass or the Earth's mass as Newton's Law of Gravitation does. The only aspect of Newton's Law of Gravitation which seems to be valid at this time is the inverse-square law nature of gravity. Therefore, since the Earth's pull equals the Moon's pull at the neutral point, the inverse-square law enables the pull of gravity at the Moon's surface to be determined. The technical derivation is presented in Appendix B. The result is that the Moon's surface gravity is 64 percent of the Earth's surface gravity, not the one-sixth or

The Neutral Point Discrepancy 49

16.7-percent value predicted by Newton's Law of Universal Gravitation!

When the reader stops to consider that the 43,495-mile figure represents the measured value of the neutral point distance supplied to us by official sources, an annoying paradox arises: Why would experts release this information and continue to refer to the Moon's one-sixth gravity condition, ignoring all the pre-Apollo references to the neutral point distance of less than 25,000 miles?

Additional information to come suggests that the Moon's gravity might even be higher than 64 percent of Earth's. In consideration of what appears to be a cover-up, and the sensitivity of the neutral point distance to slight variations in lunar gravity, NASA could have easily given the public understated figures. If the neutral point is 43,495 miles from the Moon, the surface gravity is 64 percent of Earth's. Shifting the neutral point out 8,500 miles to around 52,000 miles from the Moon has the effect of increasing the Moon's surface gravity to the same value as Earth's.

The discrepancies to be discussed in Chapter 4 involve the orbital period of spaceships around the Moon and velocities attained by spaceships reaching the Moon from the neutral point. The publicized period and velocity values are not supportive of a 43,495-mile neutral point distance from the Moon. They support the old neutral point distances and the Moon's weak one-sixth gravity. Therefore, official information is inconsistent and contradictory, indicating a cover-up. The question is why the real neutral point distance leaked out. Did some of the NASA people try to sabotage the cover-up?

CHAPTER 4

DISCREPANCIES IN SPACECRAFT VELOCITIES, FLIGHT TIMES, AND FUEL REQUIREMENTS

    Writers covering the Apollo missions consistently reported that the spacecraft were traveling less than 6,000 miles per hour when they reached the Moon. This was after they had passed the neutral point traveling slightly more than 2,000 miles per hour. The author has not seen calculations by the above writers which derive the velocity attained at the time the spacecraft reached the Moon. Therefore, it is reasonable to assume that NASA directly or indirectly supplied the information. A glaring inconsistency arises when the facts and figures are examined.

    To begin with, the 43,495-mile neutral point distance and one-sixth gravity are incompatible as explained in Chapters 2 and 3. In *Space Frontier* written in 1971, Wernher von Braun gave the velocities of Apollo 8 at the neutral point and when it reached the moon as 2,200 miles per hour and 5,700 miles per hour, respectively.[1] In the same discussion he mentioned that lunar gravity took over at a point 38,900 miles from the Moon, where the craft began speeding up again. In working out the mathematics of velocity in going from the neutral point to the Moon, the only way the craft could reach the

Moon traveling less than 6,000 miles per hour would be if one-sixth gravity is assumed. If one-sixth gravity is assumed, the neutral point is around 24,000 miles from the Moon, not 38,900. Therefore, the craft would continue to lose velocity in its coast from the Earth until it reached a point about 24,000 miles from the Moon. But this is not what the above von Braun reference said. Either the greater neutral point distance is correct with its high lunar gravity ramifications or the greater neutral point distance is wrong and one-sixth gravity exists on the Moon. Why did Wernher von Braun publish this conflicting information?

The following information is given to show the ramifications of a high lunar gravity on the velocity when reaching the Moon. One-sixth lunar gravity would accelerate the craft to a velocity of somewhat less than 6,000 miles per hour, while a 64-percent lunar gravity would boost the final velocity considerably higher. In Appendix C, the derivation of final velocity is given using the 64-percent lunar gravity required by the 43,495-mile neutral point distance. This also assumes that the initial velocity at that point is the 2,200 miles per hour claimed. This yields a final velocity of over 10,000 miles per hour! The discrepancy between the velocity demanded by a 43,495-mile neutral point, hence a high lunar gravity, and the claimed value of 5,700 miles per hour is over 4,000 miles per hour.

Before discussing orbital velocity, an argument will be presented which clearly shows that the neutral point distance is in the vicinity of the 40,000-mile mark. On the Apollo 8 flight, the spacecraft reached the neutral point, allegedly 38,900 miles from the Moon, traveling 2,200 miles per hour at 55 hours 39 minutes into the mission. At 68 hours 57 minutes, it had reached the Moon

Discrepancies in Spacecraft Requirements 53

traveling less than 6,000 miles per hour. Therefore, the distance was covered in 13 hours 18 minutes. If the neutral point were really 24,000 miles from the Moon, the ship's average velocity would have been about 2,441 miles per hour and the trip would only have taken about 9 hours 50 minutes. The times given by NASA support the greater neutral point distance, hence the Moon's high gravity.

A detailed analysis of flight times provides additional confirmation of the Moon's high gravity. If the Moon had one-sixth of the Earth's surface gravity, the Apollo 8 spacecraft would have continued to lose velocity until it reached the 24,000-mile mark. At that point, it would have begun to accelerate, reaching a final velocity of about 5,540 miles per hour at the Moon. If it were traveling 2,200 miles per hour 38,900 miles from the Moon, its time of flight would have been 16 hours 44 minutes with the assumption of one-sixth gravity. This presents a discrepancy of over three hours from the NASA-reported time of 13 hours 18 minutes. The only way to account for the shorter flight time claimed by NASA is to assign a higher average velocity and final velocity to the spacecraft. With the assumption of a 64-percent lunar gravity, the flight time is calculated to be 13 hours 47 minutes, which is very close to the NASA claim of 13 hours 18 minutes. The analysis definitely shows that the NASA-supplied information contradicts itself. The flight times and neutral point distances are indicative of a high lunar gravity, yet NASA continues to publicize the weak one-sixth gravity condition on the Moon.

If the Moon had one-sixth of the Earth's gravity, a satellite or spacecraft orbiting the Moon would have a very low orbital velocity. This is because the orbital velocity offsets the pull of gravity. If the gravity pull is less, the velocity required to maintain orbit is less. In other words, the tendency to fall is reduced, so the satellite can continue to orbit with less speed. At one-sixth gravity, a satellite orbiting the Moon at an altitude of 70 miles would only be traveling 3,655 miles per hour. However, with a lunar gravity 64 percent of Earth's, the orbital velocity at the same height would be 7,163 miles per hour, or nearly double the value claimed. In the *Encyclopaedia Britannica* within "Space Exploration", Apollo 11 reached the Moon traveling 5,225 miles per hour and had to reduce its velocity to 3,680 miles per hour to assume an elliptical orbit.[2] With a 64-percent lunar gravity, the spaceship would drop like a rock with such a low velocity and soon crash into the Moon.

It is evident that with a high lunar gravity, the braking maneuver of the Apollo crafts would only have to reduce the 10,000-plus mile per hour velocity to 7,163 miles per hour to assume orbit. This orbital velocity indicates that the craft would circle the Moon every 60 minutes instead of one orbit every two hours. The knowledge of these orbital periods must have been known by the Mission Control people because during a certain percentage of every orbit, communication with the orbiting Command Module, or CM, is cut off when the craft passes behind the Moon. The communication blackout allegedly lasted for 50 minutes out of every 120-minute orbit. In Appendix D, the velocity and communication blackout time are derived based on the 70-mile-high orbit. With a 64-percent gravity, the blackout time would only be 24 minutes.

If the above situation occurred, fairly tight security measures must have been employed to suppress information from the public. As long as the only means to monitor the astronauts' activities were through Mission

Discrepancies in Spacecraft Requirements 55

Control, only a relatively small number of people would actually be aware of the situation. The vast majority of NASA people could still be kept in the dark. If such were the case, the tightest security area would be Mission Control and the information presented below indicates this.

John Noble Wilford covered the Apollo missions for *The New York Times.* In *We Reach the Moon,* he discussed the Apollo program in considerable detail, along with his experiences in working with NASA. The following information was derived from his book.[3] Regarding the fire that killed Grissom, Chaffee, and White, he mentioned that Houston used the word gruesome to describe the tape recordings in a phone call to NASA headquarters. However, NASA officials would not admit that they knew about the tape evidence until after Wilford wrote a story in *The New York Times* on Tuesday, January 31. This incident was symptomatic of the lack of candor which the public and Congress later found NASA to be guilty of. During the weekend of the fire, reporters constantly received vague and evasive replies to their questions which made them remember Gemini 8. On Gemini 8 the capsule went out of control and the communication tapes made during the crisis were withheld because NASA believed that the astronauts' voice level might have given a false impression of their behavior. The tapes were later released and assuming that they were complete, the astronauts had handled themselves with great control. Reporters subsequently began to refer to the NASA initials as meaning Never A Straight Answer.

In *The Voyages of Apollo: The Exploration of the Moon,* Richard Lewis explained what the situation was at Mission Control during Apollo 12.[4] The following is a summary of his account. At midnight, the glass viewing room behind the Mission Control consoles was filled with VIPs such as Administrator Paine; his deputy, George M. Lowe; the astronauts Armstrong, Aldrin, and Borman; C. Stark Draper, director of the Instrumentation Laboratory of MIT where the Apollo inertial guidance system was developed; and Wernher von Braun. However, not a single member of the news media was present. Newsmen had not been allowed in Mission Control since the policy was established in Project Mercury, supposedly to prevent uncontrolled disclosure in the event of a disaster. This continued until late in Apollo when a press pool representative was finally admitted to the viewing chamber at the Johnson Space Center.

In light of information presented thus far, it seems more than coincidental that such tight security measures were employed. Clearly, the public received carefully chosen information, some of which was true, and a great deal of which must have been totally fabricated, based on the old concept of the Moon's one-sixth gravity.

As the discrepancies encountered on the way to the Moon are outlined, the discussion has brought the spacecraft into lunar orbit. The next step is to consider the ramifications of a high lunar gravity on the fuel requirements of the lunar descent and ascent vehicles. This opens up a new chamber of horrors.

To escape from the surface of a planet into space or to orbit around it, a space vehicle must be lifted to orbital height and be traveling a certain minimum speed. This requires energy to overcome the continuous pull of the gravitational field and to increase the vehicle's kinetic energy. The reader may recall that the Apollo

Discrepancies in Spacecraft Requirements 57

launch rocket which sent men to the Moon stood 363 feet tall and weighed 6,400,000 pounds. A picture of the Apollo 11 rocket at launch is shown in one of the cover photos. It was designed to send a payload of approximately 100,000 pounds toward the Moon at a velocity of 24,300 miles per hour. The Apollo 4 rocket placed 278,699 pounds into a 110-mile-high circular orbit around the Earth. A much higher payload can be placed in orbit than can be sent to the Moon because of the additional fuel and velocity needed to escape from the Earth. If the payload weight is divided into the total weight of the rocket, the payload ratio is determined. In the case of Apollo 4, this ratio equals $\frac{6,400,000}{278,699}$ or 23 to 1. This means that a launch weight of 23 times the payload was required to place the object into Earth orbit and approximately 90 percent of the rocket's weight was fuel.

If the Moon had only one-sixth of the Earth's gravity, a much lower payload ratio than the above would have been needed by the Lunar Module to soft-land or escape from the Moon. NASA claimed that the Lunar Module, or LM, weighed 33,200 pounds when full of fuel. This comprised the ascent and descent stages. A picture of the Apollo 16 LM, consisting of both the ascent and descent stages as it rested on the Moon, is shown in Photo 2. Since the loaded ascent stage weight 10,600 pounds and the empty descent stage weighed 4,500 pounds, the total payload for soft-landing was 15,100 pounds. Therefore, the payload ratio for soft-landing was $\frac{33,200}{15,100}$ or 2.2 to 1.

The loaded ascent stage supposedly weighed 10,600 pounds when full and 4,800 pounds when empty. The ascent payload ratio would then be $\frac{10,600}{4,800}$ or 2.2 to 1 also. The weights given for the full and empty LM are consistent with the fuel requirements needed if the Moon had only one-sixth gravity. Even the sizes of the tanks to hold the fuel are reasonable, hence the overall volume of the spacecraft fits the requirements for a low lunar gravity.

If the neutral point distance were 24,000 miles from the Moon, then one-sixth gravity would be expected and the fuel requirements would have been met. The astronauts could have landed with the LM and taken off again, completing their Moon exploration as planned. However, the 43,495-mile neutral point distance and its implication of a high lunar gravity remain. With the information that the Moon's gravity must be at least 64 percent of Earth's, the LM fuel requirements were calculated in Appendix E. This lunar gravity figure implies that the required payload ratio for landing and take-off must be at least 7.2 to 1. The orbital velocity to be reached or braked is approximately twice what it would be under one-sixth gravity. This requires about four times as much braking or ascent fuel.

The additional fuel requirements under a high lunar gravity become horrendous. First, the ascent stage would have to weigh 7.2 times the empty weight, or 34,560 pounds. Second, the fuel required to soft-land the fully loaded ascent stage would increase the Lunar Module's total weight to approximately 250,000 pounds. Therefore, the LM would be nearly as large as the Titan 2 rocket which weighed 330,000 pounds and was 103 feet tall! The LM supposedly weighed 33,200 pounds; hence, its weight and volume would have to increase by more than a factor of seven. The startling con-

Discrepancies in Spacecraft Requirements 59

clusion is that if men really landed on the Moon in high lunar gravity conditions, it was not done with rockets! Again, the reader must be reminded that a 43,495-mile neutral point distance implies a lunar gravity equal to 64 percent of Earth's. A 64-percent lunar gravity in turn requires a large rocket just to escape from the Moon, let alone soft-land the take-off rocket in the first place.

The problem becomes astounding if the Moon has a gravity equal to Earth's. Evidence will be presented later which suggests that this may be the case. From Appendix E, with a lunar gravity equal to Earth's, the payload ratio would have to be 18.2! This would require an 87,360-pound ascent stage alone. The descent rocket would weigh a staggering 1,589,000 pounds, one-fourth the size of the Saturn launch rocket! The Saturn launch vehicle would then need to weigh 64 times this or 101,700,000 pounds. This is about 16 times larger than it actually was.

The above enigmas bring up some interesting questions. Why did the Russians apparently pull out of the space race when they were hot on the trail of putting a man on the Moon? How did the United States succeed when rockets would clearly not work in the high lunar gravity conditions? What was the military's involvement in top secret research which led to the successful Moon landings? These questions will be answered in the chapters to come.

Until all of the evidence is presented, the case for the massive cover-up is incomplete and many questions remain unanswered. Until every aspect of the Apollo project is carefully scrutinized, the reader should maintain an open mind. After all, the public has been victimized by confidence men, politicians, militarists, scientists, and corporations for ages. Convincing stories are told, but little sound evidence is provided to substantiate the claims.

If the Moon has a high gravity, astronauts wouldn't be able to perform as expected under one-sixth gravity. Their display of jumping ability wouldn't even come close to anticipated results. In Chapter 5, astronaut athletic feats to be expected in one-sixth gravity conditions will be analyzed and then compared with how astronauts actually performed on the Moon.

CHAPTER 5

ASTRONAUT EXPERIENCES ON THE MOON

In one-sixth gravity everything would weigh one-sixth, or 16.7 percent, of its Earth weight. A 180-pound man would weigh a mere 30 pounds. Writers were speculating on the athletic abilities of men on the Moon long before the space program and Apollo. They based their calculations on one-sixth gravity. The public was anticipating some of these spectacular athletic feats when astronauts explored the Moon, but none were ever performed. The reader may remember the televised pictures of astronauts moving around on the Moon's surface. If so, the author challenges the reader to recall any extraordinary feats. In actuality, there were none.

In the November 1967 issue of *Science Digest,* an article appeared by James R. Berry entitled, "How to Walk on the Moon."[1] In it, Berry predicted that men would be able to make 14-foot slow-motion leaps, perform backflips and other gymnastics like professionals, and be able to easily move up ladders and poles with their arms.

Another prediction was given in 1969 by the Writers of *U.S. News & World Report* in *U.S. on the Moon:*

With gravity on the moon only one-sixth as strong as it is on earth, a home-run hitter in a lunar baseball game could drive a ball well over half a mile. A golfer's drive from the tee would sail clear over the horizon.[2]

The height an object will rise in a gravitational field depends on its initial velocity. If an object had the same initial velocity in one-sixth gravity as it had on Earth, it would rise six times as high. If the initial velocity of the object were doubled over its Earth velocity, it would rise 24 times as high, and if tripled, it would rise 54 times the Earth height.

A man jumps vertically by bending his kness and then flexing his thigh muscles to full extension. This propels him off the ground with a given initial velocity. If an astronaut were to jump vertically in one-sixth gravity with the same effort expended on Earth, the initial velocity would be greater than on Earth. Therefore, the astronaut would go more than six times higher.

For the purposes of this presentation, a conservative approach is taken in determining the relative jumping ability of astronauts in one-sixth gravity versus Earth gravity. A complicating factor is the alleged weight of the spacesuits and backpacks worn by the astronauts. NASA claimed that the gear weighed 185 pounds. This is a horrendous weight to carry on the Earth, but would be no problem in one-sixth gravity. Assuming that the astronauts weighed 185 pounds and their gear weighed the same, the total combined weight in one-sixth gravity would be only 62 pounds. This is still only one-third of an astronaut's Earth body weight. Therefore, the astronauts should have been able to jump vertically far higher than they could on Earth without any burden.

A number of professional athletes can jump over three feet off the ground when they are stretched out, such as in a basketball lay-up. These athletes are the exception, but an average man in good condition can easily manage 18 inches in a standing vertical jump. It will be assumed that the astronauts were capable of attain-

Astronaut Experiences on the Moon 63

ing this on Earth with a moderate effort. Finally, since John Young's vertical jumping during the Apollo 16 mission has been observed on film many times by the author, the question of spacesuit mobility and height attained can be discussed.

A standing vertical jump of at least 18 inches on Earth can be accomplished by exerting an upward force of around 500 pounds by a 185-pound person. An equation is derived in Appendix F which gives the relative heights attained by an astronaut in one-sixth gravity, carrying a burden equal to his weight, and the same astronaut on Earth without a burden. It was assumed that in each case the upward forces were identical. Since a jump from a standing vertical position only requires the knees to bend slightly, the spacesuits would not have hampered the astronauts appreciably. The televised pictures of John Young on the Moon indicated that he was able to utilize his arms and legs for jumping in an essentially normal manner.

The resulting ratio of relative jumping ability calculated in Appendix F turns out to be over four. This means that even with the astronaut gear, Young should have been able to jump over six feet off the ground if the Moon had one-sixth of the Earth's gravity. In actuality, his efforts lifted him at most 18 inches off the ground. The author's observations indicated that Young made several attempts to jump as high as he could but with no success in achieving a height of more than 18 inches. Young is shown at the peak of one of his jumps in Photo 2. Note the position of the top of Young's helmet in relation to the flag. Critics may claim that he wasn't really trying, that he purposely kept his jumping efforts to a minimum. However, if this were the case, a similar Earth effort without a backpack and suit would have lifted him only five inches or so. This was probably Young's last chance to demonstrate that the Moon had a low surface gravity. Why wouldn't he make a reasonable effort to impress the world by jumping four feet? A reasonable jump would have been conclusive proof that the Moon had a low gravity, and the risk of injury would have been minimal in low gravity conditions, even with the backpack and suit.

With the knowledge that the astronauts could only jump about 18 inches on the Moon, and assuming that the gear weighed what NASA claimed, Moon gravity is conservatively calculated to be 50 percent of Earth gravity in Appendix G. If NASA overstated the true weight of the astronauts' gear, Moon gravity would be appreciably higher. Evidence to follow suggests that the astronauts' gear weighed no more than about 75 pounds. In Appendix H, the Moon's gravity was calculated to be 71 percent of the Earth's based on the following assumptions: John Young jumped 18 inches on the Moon; his spacesuit and backpack weighed 75 pounds on Earth; and he could manage 18 inches on the Earth without any burden.

A lot of writers seemed to give the impression that the Moon suits worn by the Apollo 11 astronauts were extremely restrictive. Yet, the following information taken from *We Reach the Moon* by Wilford indicates that this was not necessarily the case.[3] Wilford mentioned that Neil Armstrong found that he could move around easily in his bulky spacesuit and heavy backpack under a lunar gravity one-sixth of Earth's. The costume weighed 185 Earth pounds and was flexible enough so that the wearer could walk, dig, climb, and place instruments on the Moon's surface. Wilford also indicated that the astronauts did not find walking and working on the

Astronaut Experiences on the Moon 65

Moon as taxing as had been forecast, and that they bounded about easily in kangaroo hops.

The idea of one-sixth gravity presents a problem in explaining how the astronauts really performed compared to how they should have performed. The difficulty in jumping cannot be attributed to spacesuit bulkiness. However, a substantial lunar gravity would create problems.

In view of the information presented thus far, it may come as no surprise to the reader that security control extended to the astronauts' conversations on the Moon as well as to Mission Control. The ability to delete and edit undesirable comments made by the astronauts could always be accomplished before transmission to the public. There was a delay from the time Mission Control received the information until transmission to our television sets.

The following is a summary of information taken from *The Voyages of Apollo* by Lewis which points out the degree of control exercised over the Apollo mission activities.[4] He indicated that the astronauts' tasks were all carefully plotted out in advance. The explorers were expected to follow the plot as faithfully as actors in a play to stay on schedule. Every move was planned, timed, recorded, and every deviation from the plan had to be explained and justified. Virtually every event and movement was governed by the flight plan—a script as large as a telephone book.

It seems that even the dialogue was carefully controlled, especially when the astronauts knew they were being filmed or recorded for television. This will be demonstrated later when references to a "hot mike" are made by one of the astronauts.

Apollo 12 was a more extensive mission than the first

Moon landing. Whereas Armstrong and Aldrin spent only 2½ hours on the Moon, Conrad and Bean would spend a total of more than 7 hours venturing a half-mile from the spacecraft. This mission was to involve many scientific experiments, including an aluminum foil solar wind collector that will be discussed in the chapter on atmosphere.

The first Apollo 12 discrepancy of significance revealing a high lunar gravity occurred just after Conrad jumped the final three feet from the bottom of the ladder to the Moon's surface. The following information was summarized from an account of the incident by Lewis.[5] As Conrad stood on the Lunar Module landing pad, he stated that the last step may have been a small step for Neil, but was a long one for him. He then stepped off the pad and mentioned that he could walk pretty well, but that he had to take it easy and watch what he was doing. As Conrad was scooping up the contingency sample, Bean warned him not to fall over since he appeared to be leaning forward too far. Supposedly, it would be difficult for him to get up in the Moon suit if he fell over. Conrad then stated that he did not think Bean would be able to steam around as fast as he thought he could.

In the above incident, it seems that Conrad was commenting on the final 3-foot jump since he referred to Neil Armstrong's jump down to the surface, not an intermediate step to the lowest rung on the ladder. Jumping from a 3-foot height in one-sixth gravity would be like jumping from 6 inches on the Earth. Even with the heavy backpack life-support systems on, the 3-foot drop would have scarcely been felt by the astronauts. They should have been able to lower themselves down with

Astronaut Experiences on the Moon

their arm strength alone, and no difficulties should have been encountered.

When Conrad began to move around on the surface, he may have experienced weight problems. However, even with the alleged weight of the gear, the astronauts should have had no problems in standing up if they fell down in one-sixth gravity. They would have been able to provide the necessary push to right themselves with their arm strength alone since their Moon weight should have only been 60 pounds or so. The evidence presented does not support the contention of one-sixth gravity. It indicates a lunar gravity close to that found on the Earth's surface.

A photo appeared in the December 12, 1969 issue of *Life* magazine showing Apollo 12 astronaut Alan Bean carrying a barbell-shaped package of instruments which allegedly weighed 190 Earth pounds.[6] The accompanying statement that it had a Moon weight of only 30 pounds does not seem consistent with the photo which shows a noticeable bow in the approximately 1-inch bar. This picture is shown in Photo 3. The movie film of this event is even more revealing. As Bean carried the instrument package across the lunar surface, the bar bent up and down, strained by the heavy burden on each end. It was also apparent that the instrument package was quite heavy from Bean's efforts and movements.

Before the remaining Apollo missions are discussed, it is of value to examine how the astronauts were given training to prepare them for their excursions on the Moon. If a 185-pound astronaut carried a backpack life-support system and spacesuit weighing 185 pounds, the combined total weight of astronaut and gear would be 370 pounds on Earth, compared to 62 pounds in one-sixth gravity conditions. Therefore, an Earth simulation of one-sixth gravity would have to lighten the astronaut and his equipment to one-third of his normal Earth body weight. Any attempt to simulate one-sixth gravity on Earth would have to be made underwater or with a special contraption which actually helps to lighten the astronaut and his burden, regardless of whether he is moving up or down. Both of these methods were employed by NASA. However, in early 1964, space scientists discovered Oregon as a place to serve as a lunar workshop without using water or special devices.

Astronauts were sent to the Bend, Oregon area to get their "Moon legs." Walter Cunningham was the first to try out the Moon suit, backpack life-support system, and certain tools to be used by Armstrong and Aldrin on the Moon excursion. In the first test over lava rocks, Cunningham lost his balance and sprained his thumb, tearing small holes in the suit glove which caused it to lose pressure. Evidently a full simulation was attempted. If so, how could the NASA people rationalize the weight problem. There is no way that the condition of one-sixth gravity could be reproduced in this manner. Even if the backpack were lightened considerably, the combined weight of a 185-pound astronaut and gear would be far more than three times the required lunar weight. If anything, the real purpose of the tests must have been to simulate a lunar gravity which is nearly the same as Earth's.

That the astronauts were able to maneuver around at all in the Bend, Oregon area with their gear on suggests that the gear weighed far less than 185 pounds. The ridiculousness of the exercise makes the NASA cover-up very clear. Since the tests began in early 1963, it is apparent that the Moon's high gravity was discovered at least as early as 1962. This supports the conclusion

Astronaut Experiences on the Moon 69

presented in Chapter 3 that Ranger Moon probes gave NASA the required information to determine lunar gravity by 1962 or earlier.

Throughout the early Apollo missions, an attempt was made to impress the public that the Apollo Moon suits were extremely bulky and awkward. This would greatly inhibit the astronauts' mobility on the Moon. Consequently, the astronauts would be effectively handicapped and incapable of impressive athletic feats. It is somewhat unbelievable that from the time Cunningham tested out the best available spacesuit gear in 1964 until the first Apollo landing in 1969, little improvement seemed to have been made in the suits. The public has always been told that the best equipment was provided for the astronauts. Certainly enough money was paid to develop the best possible equipment.

A little digging brought an interesting discovery to light. In *Suiting Up For Space* written in 1971, Lloyd Mallan stated the following:

As a matter of fact, Hamilton Standard had already achieved a space suit with 93 percent of nude range before October 1968, when they demonstrated it before the aerospace scientists and engineers attending the Fifth Annual Meeting of the American Institute of Aeronautics and Astronautics held at Philadelphia, Pennsylvania. Live demonstrations of the suit during the week-long meeting attracted wide interest and attention—plus some disbelief. It was hard for some of the onlookers to believe that so much mobility could be designed into an inflated space suit. But it was. For the advanced suit was developed to meet the greater mobility requirements of manned space missions to follow the Apollo moon-landing program.[7]

The author contends that if as early as 1968 this was the best piece of equipment available, NASA had the time and money to insure that it would be used on every Apollo mission. After all, billions of dollars were spent in sending men to the Moon. It is only reasonable to make sure that once the men are there, they can perform their tasks in the best possible manner. If they were not used, then perhaps NASA wanted to continue to convince the public that the Moon had a weak gravity. If the astronauts were encumbered, there would be less chance of a breach in the cover-up. The bulkiness and weight would be good excuses for anemic jumping and maneuvering attempts. However, it was just pointed out that the Apollo 16 astronauts had great flexibility with improved suits, yet they were still not capable of worthy jumping feats.

The public was told that modifications were made to the spacesuits by the time of the later Apollo missions. In the July 1971 issue of *National Geographic* in an article entitled "The Climb up Cone Crater", Alice J. Hall stated:

> Apollo 15's LM will be able to stay on the moon 67 hours, twice as long as *Antares* did. Improved suits will allow greater mobility as the spacemen go about their chores.[8]

The reader can compare the sizes of the Apollo 11 suits with the Apollo 16 suits and see that the latter suits were less bulky in appearance. Therefore, astronauts on the Apollo 16 mission should not have had any trouble on the Moon if one-sixth gravity conditions existed. Hills should have been climbed with leaping bounds and great distances should have been covered in short time periods by the astronauts.

Before the ill-fated Apollo 13 mission which never

Astronaut Experiences on the Moon 71

made a lunar landing, astronauts Lovell and Haise practiced a traverse in Verde Valley within Prescott National Forest, Arizona. This was to give them the experience they would need to reach Cone Crater on a ridge about 400 feet higher than the proposed landing site elevation. Again, the author suggests to the reader that practices in Arizona would be totally useless if one-sixth gravity were to be encountered on the Moon. Their centers of gravity on the Moon would be different from the Earth simulations and the Earth weights would be three or four times too high to reproduce lunar conditions. However, the practice sessions would certainly have been useful in simulating near-Earth gravity conditions.

If Cunningham's backpack and spacesuit had weighed 185 pounds, he would have become totally exhausted in minutes, but he was not. Incredibly, his 1964 simulations involved spacesuit pressure. This implies that he carried oxygen and some sort of cooling system, otherwise he would have quickly passed out from heat exhaustion. All this evidence points to the conclusion that the life-support systems and spacesuits were light enough for the astronauts to have performed in high lunar gravity conditions for extended periods of time. In addition, this was accomplished as early as 1964 and developmental efforts would have lightened the gear considerably by 1969. The combined spacesuit and life-support system weight was probably less than 75 pounds. Exotic light metals and the best known materials available to NASA would have assured this.

Following the Apollo 13 mishap on the way to the Moon, a 10-month delay was taken to re-engineer and modify the spacecraft before Apollo 14. This mission would be another attempt to reach the highland regions of Fra Mauro and the highlight of the trip was to be the

1.8-mile excursion to Cone Crater. Problems arose because the trip was mostly uphill and the astronauts had to take turns with the Modularized Equipment Transporter, or MET. On their first EVA or Moon excursion, Lewis mentioned that Shepard and Mitchell moved around with dancing steps and kangaroo jumps.[9] Unfortunately, it seems that the first excursion must have gotten the best of them because on the trip to Cone Crater, the explorers were huffing and puffing and their heart rates climbed.[10] The difficulties were attributed to their semirigid, cumbersome suits and the heavy backpack life-support systems which supposedly weighed 185 pounds on Earth.

It is important for the reader to understand that the combined weight of astronaut, spacesuit, and life-support system could not have exceeded 62 pounds in one-sixth gravity. This could hardly be considered a sizable fraction of their Earth weight. For men who were moving with dancing steps and kangaroo jumps the day before, slight hills seemed to present a formidable challenge. If the Moon's weak gravity presented such an awesome challenge to astronauts in walking uphill, then perhaps the excellent physical condition which these men were supposed to be in was overrated. It was expected to hear comments by the astronauts on the ease of moving up hills and in traversing long distances with little effort and great speed. Fortunately, Apollo missions 15, 16, and 17 did not subject astronauts quite as much to the Moon's hostile environment and tremendous "one-sixth" gravity. The Lunar Rover was to transport them most of the way to their destinations.

When the Apollo 14 astronauts were in view of the south flank of Cone Crater, Shepard went down on one

Astronaut Experiences on the Moon   73

knee to pick up a rock and required the aid of Mitchell to stand up. About two-thirds of the way to their destination, their heart rates were up to 120 beats per minute as they moved uphill. The following information was summarized from Lewis' account of the journey.[11] Their heavy breathing was heard in Houston, New York, Washington, and Florida. As they continued, the going became more and more difficult. The rim of the crater seemed close, but they could not make any significant progress toward it. As they climbed, Shepard's rate reached 150 per minute and Mitchell's went to 128. Frequent rest breaks were taken. After spending more than half of their 4-hour EVA, Shepard estimated that the rim of Cone Crater was still 30 minutes away. Shepard then concluded that there was not sufficient time left to reach the rim, even with a 30-minute extension. The astronauts never reached Cone Crater. They went back downhill toward Weird Crater to collect rock samples, then on to Triplet to dig trenches.

It seems that the 1.8-mile trek could not be negotiated. If the astronauts had not been trying so hard to reach it, this would not be very startling. After all, they had documentation and sampling work to do along the way. On the Earth, this would have been a reasonable amount of time, all things considered. But on the Moon with one-sixth gravity, the astronauts should have been able to maintain a speed of at least five miles per hour. If they were two-thirds of the way to their destination, they should have been able to travel the remaining half-mile in six minutes at a rate of five miles per hour. Yet they estimated that they could not do it in 30 minutes. If they were on Earth, they may have been able to crawl the remaining distance and still meet the deadline, but this was on the Moon, supposedly in one-sixth gravity conditions.

Finally, they returned to the landing site, checked the ALSEP instruments, and then Shepard performed his famous six-iron golf experiment. The objective of the demonstration was to show how far a ball would travel in the Moon's weak gravity. One ball supposedly traveled 100 yards and another 400 yards. The uncertainties involved preclude conclusions concerning the distances given. However, the evidence of a high Moon gravity given thus far indicates that significant distances could not have been achieved in the attempts.

The author observed one of the Apollo 14 astronauts in a movie film of the mission. The astronaut was running in semislow-motion in an otherwise perfectly normal manner. The discrepancy arises when it is considered that the astronaut went no higher off the surface and went no farther with each step than he would have on Earth. The slow-motion effects could not cover up this fact. This suggests that the film speed was adjusted to slow down the action to give the impression that the astronauts were lighter than they actually were. With the slow-motion effects, objects would appear to fall more slowly and the public would be convinced of the Moon's weak gravity.

In a 1979 television special commemorating the tenth anniversary of Apollo 11, a short replay of astronauts on the Moon was given. The author had hoped to see many minutes of televised pictures from the Moon. However, the hour-long special allocated less than two minutes of time to the films. To make the situation worse, it appeared that the film was edited to remove many frames between pictures. This made the film choppy and the astronauts appeared to move around at super-high speed in the style of old-time movies. Perhaps other viewers asked themselves why such poor coverage and little attention was given to the original films of this historic event. Instead, the special focused on preparations for the trip and other aspects of the astronauts' lives. Alan Shepard commentated and made a point to mention the one-sixth gravity condition on the Moon.

Apollo 15 was to employ the Lunar Rover land vehicle for the first time to enable astronauts to cover greater distances. After the problems Mitchell and Shepard had on the Moon, this was almost a necessity. The Apollo 15 mission was to the Mt. Hadley-Apennine Mountain region where the Rover would take them up fairly steep slopes. Even with the Rover, Scott and Irwin had to stay within a six-mile radius of the Lunar Module. This was the maximum walking distance back to the ship in the event of a breakdown.

The Rover was supposedly designed for the Moon's one-sixth gravity, but close examination indicates that it resembled a vehicle more suitable for near-Earth gravity. It was approximately 10 feet long and 4 feet high, with a 7.5-foot wheelbase and 6-foot tread width. The wheels were 32 inches in diameter with chevron-shaped treads of titanium, not much different looking than an Earth tire. Each wheel had its own quarter-horse electric motor and its top speed was given as 17 kilometers per hour, or 10.6 miles per hour, on the Apollo 16 mission. It had an Earth weight of 460 pounds and could carry a load of 1,080 Earth pounds. A picture of Irwin and the Rover is shown in Photo 4 with Mt. Hadley in the background.

On the Moon, assuming one-sixth gravity, the empty Rover with scientific and communication equipment would weigh less than 120 pounds. The astronauts had to unload the Rover from the side of the Lunar Module and unfold it before they could use it. According to Lewis, it was harder for them to unload the Rover on the Moon than it had been in practice sessions on Earth.[12] During the process, numerous phrases such as "take it easy", "atta boy", and "easy now" were heard in the conversations between the Earth and the Moon. It seems that two astronauts were struggling with an object which should have only weighed 120 pounds or less in one-sixth gravity. The deployment of the Rover had been practiced on Earth and should not have been harder on the Moon. It should have been easier, but it was not. It is significant that Scott and Irwin wore updated Moon suits with a convoluted neck and waist.[13]. This enabled them to turn, nod their heads, twist, and bend forward a lot more easily than prior astronauts. Moon suit bulkiness seems to be a poor excuse for the astronauts' performance in unloading the Rover. The only remaining possibility is the high lunar gravity since practice sessions were conducted to preclude any mechanical problems or procedural difficulties.

The problems anticipated on the Moon dealing with lunar surface vehicles were discussed in a 1966 book entitled *Survival on the Moon* by Lawrence Maisak.[14] The writer stated that stability would be one of the most vexing problems due to the weak gravity. He mentioned that the center of gravity should be kept low and the wheel tread kept wide to prevent the vehicle from overturning. Accordingly, a minimum-size vehicle would need a wheelbase of 20 feet to give it speed capability over rough terrain. However, a much longer wheelbase would restrict the fore and aft obstacle clearance. The body of Maisak's vehicle would be a 7-foot diameter cylinder having 3 feet of clearance when above level ground. To keep the center of gravity to within 6 feet of the surface, it would need a tread width of 20 feet. Maisak was attempting to design a vehicle for rough terrain in one-sixth gravity. His proposed design ensured that the vehicle would clear rocks and still maintain stability in gravity conditions which could easily overturn an Earth-type vehicle.

An analysis is presented in Appendix I to determine how the Rover would have performed if the Moon had one-sixth gravity. The Rover encountered mostly loose dust and rocks on the Moon. This type of surface would have less traction than ordinary pavement. The Rover had a loaded Earth weight of 1,540 pounds. Under one-sixth gravity, only 128 pounds of force would be required to make the vehicle slide. Therefore in going the maximum speed of 10.2 miles per hour, the vehicle would begin to slide if the wheels were turned enough to make a radius of curvature of less than 84 feet. Even at 5 miles per hour, the minimum curvature would be 20 feet. The operator would have to be extremely careful not to make any abrupt changes in direction since a sharp turn could tip it over. The Lunar Rover would be especially dangerous because the astronauts were supposedly carrying the heavy backpack life-support systems which extended about 5 feet above the lunar surface. The vehicle seats were just about 3 feet above the ground; therefore, the majority of the astronauts' combined weight of 800 Earth pounds was well above this. Thus, the Lunar Rover does not follow the guidelines for lunar vehicle design proposed by Lawrence Maisak in the above cited reference.

The maximum braking force that could be exerted also depends on the vehicle's lunar weight. Under one-sixth gravity, only 128 pounds of braking force would be exerted by the locked wheels. This could only slow it down 2.68 feet per second every second. At this rate, it would take almost 6 seconds and 42 feet to stop the Rover going 10.2 miles per hour. This would be acceptable on a flat surface with no obstacles; but on the Moon, rocks and ruts of appreciable size could not be avoided in time to keep from damaging the Rover or tipping it over. It isn't difficult to see that the Rover would be a dangerous vehicle to drive on the Moon in one-sixth gravity. Climbing and descending steep hills with the Rover would be like committing suicide if one-sixth gravity conditions existed. This provides convincing evidence that the Moon has a high surface gravity nearly equal to Earth's.

In Apollo 16, still more modifications were made to the Moon suits. NASA delayed the launch until March 17, 1972 to strengthen the more flexible suits and to rework a docking jettison device to ensure a clean separation during lift-off.[15] Apollo 16 provided valuable information in regard to the real findings of the space program. The reader will recall the jumping feats of John Young referred to at the beginning of this chapter. He was performing with the improved Moon suit which was supposed to be even better than the Apollo 15 improved version. One might even suspect that this latest suit measured up to the Hamilton Standard design first exhibited in 1968.

On their first EVA, Young and Duke tested out Rover II. Young drove the Rover at maximum acceleration as they approached the landing site, reaching 17 kilometers per hour. According to Lewis, the surface was rough and they wanted to see how the vehicle would perform in "Grand Prix" driving while making sharp turns with speed.[16] Clearly, the exhibition would have met with disaster in one-sixth gravity conditions.

At the end of Apollo 16's first day on the Moon, the "hot mike" issue came up. It was suggested previously that the astronauts were careful about what they said when they knew the conversations might be heard on the public address system. A summary of the incident derived from Lewis' account will now be presented.[17] Young and Duke evidently thought that the microphones were off and began to converse in more visceral and explicit terms than they normally would have used if they believed they were still being heard over the space center's public address system. Houston then sent up a call to Young, telling him that he had a hot mike. Young apologized and mentioned that it was sometimes a terrible thing to have a hot mike up there. Houston then told them what a commendable job they had done considering that they didn't know they were on.

From the above information, the reader can see that the astronauts were carefully monitored by Houston. In addition, they generally maintained control of themselves when their microphones were on. In the above instance, it seems that an instrument problem caused them to believe they were off the air. Information uncovered of this nature is meaningless taken as an isolated occurrence. However, in the context of the material presented in this book indicating a massive cover-up, this bit of evidence carries a lot of weight.

Charles Duke evidently had a difficult time on the Moon. He fell a number of times and a series of photos appeared in many newspapers which showed him stumbling and falling. Incredibly, these falls were actually presented by the news media as a demonstration of the Moon's weak gravity. Since objects would take nearly

2½ times longer to fall in one-sixth gravity, Duke should have had plenty of time to catch himself. It is even more surprising that Duke fell as often as he did considering that he was wearing the most advanced, updated Moon suit which supposedly provided him more flexibility than any astronaut before him.

The destination of Apollo 17 was to a valley surrounded by a mountain system, southeast of the Serenitatis Basin. Cernan and Schmitt began their first EVA by deploying and loading the Rover. The following interesting account involving Cernan was summarized from *The Voyages of Apollo*.[18] It seems that Cernan was so enthusiastic that the Capcom (Capsule Communicator), Parker, warned him that his metabolic rate was moving up. This meant that he was using more oxygen. Cernan replied that he had never felt calmer in his life and indicated to Parker that they would take it easy. He mentioned to Parker that he thought it was due to getting accustomed to handling himself in "zero G." Parker, an astronomer, then stated that he thought Cernan was working at one-sixth gravity. Cernan's reply was, "Yes. You know where we are . . . whatever." The latter remark by Cernan in response to the Moon's gravity seems to suggest that he wanted to avoid the discussion. Perhaps Parker was not aware of the high gravity situation and asked an embarrassing question.

The remainder of the Apollo 17 mission was devoted to scientific experiments. Since Schmitt was a geologist, a great deal of field study was conducted with many Moon rock samples. In addition, experiments with gravimeters, atmospheric composition detectors, and a device to determine if water or ice existed below the surface were conducted. Since atmospheric experiments were also a part of the Apollo 15 and 16 missions, it is reasonable to conclude that the density of the atmosphere was worthy of study and that the findings of prior missions indicated a further need for measuring it at different locations. If the Moon were as much of a vacuum as has been claimed by scientists, the repetitive atmospheric measurements should not have been necessary. It is interesting that the postulate of the Moon's vacuum condition is based upon the Moon's weak gravity. A substantial gravity is required to hold an atmosphere.

Keeping these concepts in mind, lunar atmospheric theory, based on the concept of one-sixth gravity, will be explained in Chapter 6. The implications of a high Moon gravity will also be given.

CHAPTER 6

MOON ATMOSPHERIC THEORY PRIOR TO THE SPACE PROGRAM

Orthodox science has always contended that the Moon is a completely airless world. The primary reason has been that the Moon's weak one-sixth gravity would be unable to hold much of an atmosphere. Any indications of a substantial atmosphere would be ignored by most orthodox scientists because they would be convinced of the weak gravity beforehand. Evidence of a high lunar gravity has already been presented. The purpose of this chapter is to give the reader a clear picture of conditions that would exist on the Moon without an atmosphere. When evidence of a significant atmosphere is presented later, the extent of the cover-up will be apparent.

The following analysis of conditions to be expected on the Moon was given in *U.S. on the Moon* by the writers of *U.S. News & World Report* in 1969:

If the moon once did produce the ingredients for an atmosphere, these would have been lost because the moon's gravity is too weak to keep oxygen, nitrogen, and the other gases that give life to earth from escaping into space. And without an atmosphere, there could be no water on the surface. . . .

Viewing the skies from the moon's surface, one might conclude that all the universe is just as barren. The stars are in view night and day, but they never twinkle because there is no disturbed atmosphere which makes them appear to do so. The vast millions of miles of space between stars are pitch black. From the moon the sun appears to be a ball of intolerably bright radiance, but the sky around it is as black as midnight.[1]

A vacuum condition on the Moon would also cause dust particles on the surface to behave much differently than on Earth. The nature of the Moon's surface in a vacuum is easily predicted by a simple experiment. The following summary of such an experiment was written from information found in *Exploration of the Moon* by Franklyn M. Branley.[2] Fred Whipple of the Smithsonian Astrophysical Observatory in Cambridge, Massachusetts contended that dust particles would become tightly packed together without gases to filter in between and separate them. Consequently, Whipple and his supporters maintained that dust is so compacted on the Moon that a strong crust capable of supporting men and their vehicles would exist. An experiment to verify this was conducted by Dwain Bowen of The North American Aviation Company. A steel ball was released into a container of fine dust-like particles and promptly sank. When the ball was dropped under the same conditions in a near vacuum, the ball stopped at the surface. The resulting crust consisted of dust particles so compacted that a semisolid was created capable of supporting the ball.

Even Wernher von Braun seemed to agree with the above logic in his 1971 book, *Space Frontier!*[3] He stated that he and many people had always held that there cannot be very much loose dust on the Moon. Von Braun mentioned that a simple experiment shows that dust in a vacuum, such as on the Moon, becomes hard-packed, and that adjacent dust particles will fuse together into a

Moon Atmospheric Theory Prior to the Space Program 85

pumice-like substance. From the information just presented, it is clear that no dust could exist in a near vacuum. If the Moon had only one-sixth of Earth's surface gravity, it could not hold an atmosphere and the surface would be nearly as hard as compacted dirt.

Another old time belief about the Moon was that it would exhibit no signs of weathering or erosion. This is because weathering and erosion effects are primarily the result of atmospheric phenomena such as rain and wind. In a vacuum there can be no clouds, rain, or atmospheric wind. Consequently, the consensus used to be that the Moon would have a rugged, jagged terrain with little or no rounding of mountains. If any crosion or weathering effects were found, they would have to be due to volcanic activity, meteoritic or micometeoritic bombardment, temperature changes, or the solar wind. The solar wind is a supersonic flow of hydrogen and helium gas from the Sun which blows continuously through the solar system.

Supposedly, no water exists on the surface because the hot lunar day would evaporate it and the weak gravity could not prevent it from escaping into space. Without air and water, surface color changes could not occur other than those produced by volcanic activity and meteors. Seasonal color changes would never occur without weather and vegetation.

The lunar day is 28 times longer than an Earth day; therefore, it takes 14 Earth days (24 hours long) for the Sun to move across the lunar sky from sunrise to sunset. The lunar night is also 28 times longer than an Earth night, equivalent to 14 Earth days. The lack of atmosphere and a day and night cycle which is 28 times longer than Earth's would cause the Moon's daily temperature to vary more than 500° F. The longer lunar day causes the surface to reach a higher temperature and the lack of an atmosphere would prevent heat from escaping as rapidly because there would be no air to carry heat away from the surface. During the night, the reverse situation would occur. The surface heat would radiate into space more quickly than if an atmosphere existed and the long night causes the temperature to drop to extremely low levels. The Earth's atmosphere acts as a heat storage bank to keep surface heat from escaping during the night and to prevent it from building up too much during the day.

Even during the day there would be great differences in Moon surface temperatures between shadows and sunlit areas. This is because the surface heat in shadows would escape rapidly while sunlit surface heat would not. This effect is noted on the Earth at high elevations and is due to the thin air.

Surface temperature differences between shaded and sunlit areas would create other problems. Material not exposed to sunlight would become brittle and shatter with little provocation due to low temperatures. High temperatures would make objects exposed to the Sun too hot to handle after a short time. Objects only partially exposed to the Sun would experience tremendous destructive thermal stresses due to temperature differences between the sunlit and shaded sides. Spaceships and scientific instruments on the Moon would have these problems if not properly shielded. If an astronaut stayed in one position too long, he might be cooked on one side and frozen on the other. Without protective clothing, the astronaut would be fried if he sat on a rock or a Moon vehicle seat exposed to the Sun.

Light diffusion could not occur without an atmosphere. Shadows could only be illuminated by reflected

Moon Atmospheric Theory Prior to the Space Program 87

light from other surfaces. Without these other sources of light, shadows and objects in these shadows would be nearly, if not completely, invisible. The Sun's disk would be clearly evident on any photo and blackness would extend to the Sun's corona. No halos would be seen around the Sun because no atmospheric diffusion (scattering) or refraction (bending of light) could occur. Sunset and sunrise effects on the Moon would never occur because diffusion and refraction through the atmosphere would be nonexistent. Without dust in the air and an atmosphere to hold it, no light scattering would take place.

Without an atmosphere, meteors would never be seen burning up above the Moon's surface. In addition, stars would never appear to be occulted by the Moon. Stars are occulted when they are eclipsed by the Moon or other planets. If the planet has an atmosphere, stars close to the surface will be dimmed and reddened due to refraction. They also appear to be displaced relative to other stars and remain visible for a little while after they have actually been covered by the planet's disk. This refraction or light bending also causes the star to reappear on the other side of the planet somewhat earlier than it would without an atmosphere. During a solar eclipse, the Sun would shine through the Moon's atmosphere if it had one. This would produce a refraction halo around the Moon.

Moon explorers should have had no problems with dust because dust cannot exist in a vacuum. If they somehow managed to create dust and coat themselves with it in this vacuum, it would be almost impossible to remove. The dust particles would cling to them like glue.

Without an atmosphere, ordinary machinery with moving parts which are not lubricated would cease to function. Layers of air molecules which are bound to the surfaces tend to prevent these surfaces from adhering or binding when they are in contact with each other. All surfaces would be very sticky.

The implications of a high lunar gravity are devastating to the old Moon vacuum theory. A high lunar gravity implies that an atmosphere exists because volatile substances and gases continuously given off by the planet cannot escape the pull of gravity. The Moon would soon reach an equilibrium state where the density of air at the surface would remain essentially constant. The air pressure would also depend on the elevation, just like the Earth. The reader should consider that evidence of a substantial atmosphere is also evidence of a high lunar gravity. Verification of one assertion verifies the other. A substantial atmosphere on the Moon means that clouds, weather, erosion, water, plant life, and animal life may exist. However, the conditions cannot be the same as on Earth because of the long lunar days and nights and other considerations.

In the next chapter, allusions to the Moon's vacuum will be shown to have about as much merit as the contention of one-sixth gravity. A startling number of breaches in the NASA cover-up supplied a great deal of evidence that the Moon has an Earth-like atmosphere. But in addition to the NASA-supplied evidence, many other sources of information exist. These will be explored in considerable detail.

CHAPTER 7

INCREDIBLE FINDINGS CONCERNING THE LUNAR ATMOSPHERE

There were many indications during the Apollo missions that the Moon has a Earth-like atmosphere. Television commentators and reporters seemed to ignore these indications and went along with the accepted belief that the Moon is airless. Evidence confuting the vacuum theory will be presented and conventional beliefs given in the previous chapter will be examined in light of the evidence.

Dust cannot exist in a vacuum; however, the reader probably remembers watching the astronauts walk through it on the Moon's surface. Just before astronaut Neil Armstrong stepped onto the Moon for the first time, he described the surface as fine grained, almost like a powder.[1] Once on the surface, Armstrong confirmed his initial observation by stating that it was fine and powdery. He described how he could pick it up loosely with his toe, and how it adhered in fine layers to the soles and sides of his boots like powdered charcoal. He mentioned that he only went in a fraction of an inch; that he could see his footprints in the fine, sandy particles.[2] Orthodox scientists accept that this dust exists, but they continue to deny that the Moon has any substantial atmosphere. There has to be an atmosphere because dust cannot exist in a vacuum. The density of the atmosphere remains to be determined.

Apollo 11 landed in the Sea of Tranquility which is a lowland on the Moon. If an atmosphere existed, its density would be greater in the lowlands than anywhere else. Since Apollo 12 also landed in the lowlying Ocean of Storms, additional atmospheric indications would be expected during this mission. Indeed, soon after the landing, Conrad stated that he thought they were in a lot dustier place than Neil Armstrong was.[3] Apollo 17 also encountered dusty conditions. The reader may recall Apollo 17 pictures of the Rover with the "rooster plume" of dust coming off the back wheels. The dust not only stretched behind the Rover, but it curved around and rained down on the astronauts.

A picture of the deployed Solar Wind Composition Experiment on the Apollo 11 mission is shown in Photo 5. This was a very thin strip of aluminum foil, four feet long and one foot wide, which was to hang straight down from a support rod. It was designed to stop the solar wind particles which would be analyzed back on Earth. The same experiment was performed on the Apollo 12 mission. The following Apollo 12 incident involving the experiment was summarized from an account given by Lewis.[4] Earlier in the day, before their second EVA, astronaut Bean noticed something through the LM window which baffled him. The solar wind trap had been hanging straight down when the astronauts entered the LM after their first EVA. Bean reported to Capcom (Capsule Communicator) that the sheet looked like a sail in the wind around the pole, with a bulge in the front and bent back on both sides. Capcom replied that he suspected they had a real solar wind. Bean replied to Capcom that Capcom must be kidding. In turn, Capcom offered an alternative explanation: The front was thermally expanding more than the back; the

Incredible Findings Concerning the Lunar Atmosphere

back was radiating and the front was hot due to a thermal difference across it. Capcom stated that he was receiving a lot of approval from others in Mission Control on the idea. Bean then insisted that it still looked like it was wrapped around the pole as if a wind were blowing on it.

After the astronauts left the LM for their second EVA, Capcom told Bean to ask Conrad to take a picture of the solar wind composition sheet as it was wrapped around the pole. Another amazing thing then happened as Conrad prepared to take the picture. The foil no longer appeared to be wrapped around the pole.[5] Conrad informed Bean that it must have been an optical illusion from inside the Lunar Module. Bean then reported this observation to Houston.

An analysis of the incident is now in order. The solar wind was expected to be substantial and measurable on the Moon's surface if an atmosphere did not exist. This vacuum condition is why scientists designed the experiment in the first place. The primary constituent particles of the solar wind are supposed to be hydrogen and helium traveling with velocities up to 1000 kilometers per second. However, the solar wind is so miniscule that its density only fluctuates between 1 and 30 hydrogen atoms per cubic centimeter during quiet Sun activity. The assumption of a solar wind density of 4 atoms per cubic centimeter yields an impact pressure of 10^{-8} dynes per square centimeter. If 10 times the pressure is assumed for an active Sun, the pressure turns out to be .000000000000034 pounds per square inch. This pressure could not even be measured without specially designed sensitive instruments. The normal atmospheric pressure on Earth at sea level is 14.7 pounds per square inch. A barely noticeable 1-mile per hour breeze on

Earth exerts a pressure of .000018 pounds per square inch and will barely move a window shade. However, this is 53,000,000,000 (53 billion) times as strong as the solar wind!

Clearly the solar wind was not responsible for causing the composition sheet to bend back against the pole. After Capcom suggested the real solar wind idea, Bean came back with a statement alluding to the joke which he thought Capcom was making. Bean evidently knew that the solar wind was not responsible. Capcom evidently knew it also, but may have been trying to patch the leak. Capcom then countered with an explanation which fell apart when the solar wind sheet mysteriously straightened out.

Capcom suggested that the front of the sheet was expanding more than the back. This, of course, happens to objects in the Sun, but other evidence destroyed this explanation. First of all, the sheet had been set up in the Sun for many hours without any noticeable effects on it. The astronauts were out working with the instrument and would have noticed the warped sheet on their first EVA. Since the Sun would heat up the sheet very rapidly if a vacuum existed, the warping would occur almost immediately. Once it became warped, it would stay that way until taken down. The Sun would move very little across the Moon's sky in the time of the trip. Therefore, the conditions creating the warped sheet in the first place would remain nearly the same.

After Capcom made the thermal expansion suggestion, he mentioned that he was getting approval of the idea from other Mission Control people. Evidently they must have thought that it would be an acceptable explanation for the public and other scientists. Bean replied that it still looked like it was wrapped around the

Incredible Finds Concerning the Lunar Atmosphere

pole with the wind blowing on it. The statement by Bean is clear. He seemed convinced that what caused the sheet to bend around the pole was a real atmospheric wind, not the solar wind. His persistence in referring to the wind after the solar wind idea had already been rejected by Capcom and himself is a giveaway. Bean seemed to ignore the explanation by Capcom and may have been fascinated by the atmospheric evidence.

The final blow to the thermal expansion explanation occurred when Conrad went to photograph the bent sheet. It had mysteriously straightened out in a short time. As explained previously, this would not happen under conditions demanded by a vacuum. The reference to an optical illusion was probably a means to end the conversation about something which should never have been brought up in the first place. Once the cat was out of the bag, Capcom and the astronauts would have to patch up the security breach as best they could. Bean was a trained astronaut and seemingly would not have been duped by an optical illusion.

If the Moon's atmosphere could produce a noticeable movement of the solar wind composition sheet, it has to be fairly dense. During some of the Apollo missions, dust kicked up by the astronauts had a tendency to drift. This is an indication of a dense atmosphere. In addition, the American flags billowed noticeably during some of the early Apollo missions. The first Apollo missions had flags with horizontal support rods along the tops to make them stand up. This still allowed light winds to cause them to bow out or billow occasionally.

The author acquired the movie showing the Apollo 14 flag ceremony. Close analysis of this film shows that the flag billowed and waved when the astronauts were not touching it or even close to it. At the end of the flag ceremony, as one of the astronauts moved away from the flag, it began to wave back and forth. In an apparent attempt to mask any further display of an atmospheric wind, both astronauts blocked the movie camera's view of the flag. The astronaut nearest the flag began running toward and in front of the camera while the other astronaut put his arm in front of the lens. However, it was already too late to cover up the evidence. The author challenges all skeptics to witness this film for themselves and to explain the waving flag by any other logical means than a dense lunar atmosphere. It is amazing that this hard core evidence is available to the public. Even a diehard skeptic should be convinced after seeing this film sequence. The Apollo 14 film was ordered in 1980 from Movie Newsreels, a company located in Hollywood, California.

On Apollo 16, there was an apparent attempt to reorient the public's thinking about the waving flag evidence already released. This time a heavily starched flag that would maintain a distorted shape at all times was opened. A special point was made on one telecast to stress that the flag was processed in this way to give the appearance of a waving flag in an airless world. The real purpose was undoubtedly to make the flag less susceptible to breezes than previously unstarched flags had been.

Photographic evidence of light diffusion is one of the best proofs of a dense lunar atmosphere. An Apollo 14 picture of the lunar surface, Lunar Module, and tire tracks from the MET (Modularized Equipment Transporter) is shown in Photo 6. It shows streamers of light across the entire lunar surface and horizon. The diffusion of light from the Sun is so great that most of the visible sky is illuminated. The reader should recall that

Incredible Findings Concerning the Lunar Atmosphere 95

references were given in Chapter 6 explaining that in a vacuum the Sun would be extremely bright, but the sky around it would be completely black. Photo 7 is a picture of Apollo 15 astronaut Scott standing on the slope of Hadley Delta with the Apennine Mountains in the background. Photo 8 shows Apollo 16 astronaut Duke scooping samples on the rim of a deep crater with the Rover in the background. The visible sky in both of these photographs is very bright, indicating the even diffusion of light through the lunar atmosphere. Clearly, the photographic evidence does not support the assertion of a lunar vacuum. It provides evidence of a dense atmosphere.

Another photo revealing the NASA cover-up appeared on the cover of the December 12, 1969 issue of *Life* magazine, showing Apollo 12 astronaut Bean setting up instruments on the Moon. He is surrounded by what appears to be a halo.[6] This picture is shown in Photo 9. Since other photos of astronauts on the Moon do not show a semblance of this halo effect, there seems to be only one reasonable conclusion: In NASA's attempts to suppress visual evidence of an atmosphere, the sky was blackened or touched up in all but a few of the photographs. The halo which appeared around Bean was the result of a poor job of masking or obliterating the sky around him. Since this lunar light effect was so pronounced in this picture, it should have appeared in other photos but it did not. Other writers speculated that it was the astronaut's aura, or radio energy emissions which became visible in the Moon's vacuum. This might have had some merit if it consistently appeared, but cannot be taken seriously in light of the other photos. Significantly, NASA never bothered to give the public a reasonable explanation of why it occurred; they simply attributed it to spurious reflections from his suit. However, this does not make any sense because the "spurious reflections" would still have be reflected off something in the space surrounding him to account for the phenomenon. In a vacuum, a camera only picks up light photons which travel in straight lines from each point in the picture; therefore, Bean should have been surrounded by a jet-black sky in the space above the horizon. The amount of reflected light surrounding him is so great that only a dense atmosphere can account for it.

An excellent example of how the blue sky was filtered out of Moon pictures is provided by the movie film of Apollo 14 astronaut Mitchell's descent down the Lunar Module ladder. As he began his descent, the amount of light diffusion from the sky was so intense that the entire sky was almost white with shades of blue. In addition, it was difficult to see specific details of Mitchell and the Lunar Module due to the amount of light. Incredibly, as he made further progress down the steps, the white and blue sky gradually turned to light blue, then to a darker blue, and finally to extremely dark blue or black by the time he reached the surface. By then, all details in the film were clearly outlined with little, if any, light diffusion evident. The next scene in the film was the previously mentioned flag ceremony which had a very dark sky. This film segment of Mitchell shows that either the camera had a filtering capability or the film was altered after it was returned to Earth. In any event, this incident provides evidence of a dense, blue lunar atmosphere in support of evidence already provided by the waving flag on the same movie film. Significantly, it provides proof that the capability existed to filter out the blue sky in Moon photographs and films. Photo 10 is a frame taken from the movie sequence showing the blue sky. Mitchell is descending the LM ladder.

Mists, clouds, and surface changes have allegedly been seen at various times on the Moon. Many such observations are referenced in an article by Paul M. Sears entitled "How Dead Is the Moon?" which appeared in the February 1950 issue of *Natural History*. The following is a partial summary of observations referenced in this article.[7] Besides twilights which demonstrate the lunar atmosphere, more spectacular proof is provided by observations of bright moving specks which might be luminous meteors in the lunar air. There are monthly appearances of strange dark areas known as variable spots which spread and grow as the Sun climbs, becoming darker in relation to the rest of the surface. Some of these spots fade again toward sunset, while others steadily darken till night prevails. These spots may differ in size and shape from month to month, and some spots even occasionally fail to reappear. Sparse clouds seem to occasionally drift over the surface and obscure surface detail. Some of these clouds are outlined on one edge by their own shadow and there are certain regions where clouds are seen more frequently than others. For instance, six astronomers in the last century claimed to have seen a mist which obscured details in the floor of the crater, Plato.

The astronomers who made the above mentioned discoveries were not taken seriously because the Moon's one-sixth gravity was not supposed to have enabled these phenomena to occur. In *Strange World of the Moon,* written in 1960, V.A. Firsoff mentioned that experienced observers have recorded changes in the intensity of dark and bright markings during lunar eclipses and other times during the lunar day. He referred to local fade-outs of visibility when the rest of the Moon is not affected. Shades and patches which resemble mists and clouds, as well as glows and lights, have all been observed to appear and reappear in certain localities. Firsoff stated that all of these phenomena cannot be solely attributed to lighting effects or the position of the Moon with respect to the Earth.[8]

According to Firsoff, an area of brightness exists in the lunar Alps where some of the peaks appear ill-defined on occasion, even though the surrounding country is sharply outlined. Furthermore, in the southeastern portion of Mare Crisium, near Picard crater, some of the obscured regions have persisted continually for years, completely covering surface detail.[9]

Firsoff referred to sunlight seen near the terminator which is redder than under a high lunar Sun. He mentioned that it is difficult to explain this in any other way than scattering by gas combined with tiny crystals. Other colors of green, brown, blue, and violet have been seen in the maria and inside craters. The periodic change in intensity, position, and extent suggests that the color effects are caused by physical or chemical changes which depend on the amount of solar heat. Firsoff even considered that biological activity might account for the observations.[10] The above considerations add credence to the NASA-supplied evidence of a substantial atmosphere.

The occultation of stars by the Moon is additional evidence for the existence of an atmosphere. Charles Fort wrote a book entitled *New Lands* in 1923 in which he discussed the numerous observations of Moon-occulted stars.[11] Apparently, the seeming motion of stars occulted by the Moon was such a commonly observed

Incredible Findings Concerning the Lunar Atmosphere

phenomenon at the time that Fort was certain that the Moon had an atmosphere. Unfortunately, the data regarding occultation which has been collected is made somewhat indeterminate by the irregular shape of the Moon's outline. In addition, the data is so variable that estimates of the density of the Moon's atmosphere cannot be relied upon. The occultation measurements only establish that there is an atmosphere, not its density.

Eclipse halos have some of the same problems that Moon-occulted stars do. An eclipse photo which appeared on the cover of the April 1979 issue of *Life* magazine may demonstrate this halo effect. However, critics might argue that the halo is made up entirely of the Sun's corona, not the Moon's atmosphere. In any event, disregarding the solar flares, the halo's thickness indicates that it becomes almost imperceptible at an altitude of 150 miles above the surface. At this height, the density of the Moon's atmosphere would be negligible. It would be so thin that the Sun's light would not interact with the rarefied air molecules to any extent.

Conclusions regarding the Moon's atmosphere have always failed to take into account the extent of dust particles and water vapor suspended in it. These particles may be the greatest factor governing the diffusion of light through an atmosphere. Firsoff mentioned that the lunar atmosphere would probably scatter light like a pure gas, unlike the Earth's atmosphere which contains a high percentage of dust, ice crystals, and water droplets. He stated that even at the altitude of Pic du Midi Observatory (9,351 feet), the Earth's atmosphere will scatter longer wavelength light a lot more than the Rayleigh gas scattering formula predicts because of the large particles in it. [12]

The Moon's atmosphere is not likely to experience high winds and other weather conditions to the extent that is common on Earth because of the long days and nights, and the absence of large bodies of water on the surface. Therefore, the atmosphere is probably much cleaner than the Earth's and light diffusion and scattering effects would be minimal. In addition, sunrise and sunset color effects would not be as great and the atmospheric halo would be less apparent at the time of a solar eclipse. Occultation of stars would not be as pronounced as expected and the interpretation of the findings would be that the Moon has an extremely thin atmosphere. The atmospheric density could still be as great or greater than Earth's without being as visible.

It is logical to assume that the Earth and Moon, and hence their atmospheres, were created in the same way. An atmosphere is probably produced by the discharge of solid and gaseous material from the upper crust. Bodies like the Earth and Moon will possess atmospheres of the same depth if their gravities are the same and sufficient to hold an atmosphere. If one of them has a lower gravity, its atmosphere will be deeper because the gases are compressed to a lesser extent than the atmosphere of the one with the stronger gravity. The depth of an atmosphere is inversely proportional to the magnitude of the gravitational field. This follows from the gas law involving pressure and volume which states that the volume of a confined gas is inversely proportional to the pressure applied to it. In other words, if the pressure is doubled, the volume becomes half as great.

If the Moon has as much atmosphere in relation to its surface area as the Earth, is there any direct evidence that indicates it? According to Paul M. Sears in the previously cited article, lunar astronomers in the 1930's

Incredible Findings Concerning the Lunar Atmosphere

began to speculate on the fate of meteors they knew must be striking the Moon. Calculations were made which showed that meteorites weighing ten pounds or more, impacting on the dark portion of the Moon (assumed to be airless), should disintegrate in a flash brilliant enough to be seen with the naked eye. Over 100 such flashes should occur each year. As a matter of fact, only two or three such flashes had been reported in all history. This meant that they were being consumed in an atmosphere before striking the surface. The Moon seemed to be better protected from meteorites than the Earth!

To explain this paradox, astronomers reasoned that although the density of the Moon's atmosphere at its surface is only 1/10,000 of the Earth's, its density above 55 miles is greater than the Earth's at the same elevation. This was attributed to the Moon's low one-sixth gravity which was unable to concentrate its atmosphere near the surface. However, if the Moon's atmospheric density were only 1/10,000 of the Earth's at the surface, and its gravity were only one-sixth of Earth's surface gravity, the mass or quantity of atmosphere over a unit area would only be 6/10,000 of that protecting the Earth. Since the quantity of air is the most important factor which protects the surface from meteorites, the paradox was not resolved by their explanation. There simply wasn't sufficient air in the lunar atmosphere to account for the burnup of meteors if one-sixth gravity was assumed.

According to the Sears' article, it was apparent that the brightest meteors, those that would be reported as fireballs if they occurred on Earth, should be faintly visible through telescopes. Therefore, in 1941, one of the most experienced students of the Moon, Walter

Haas, began an extended search for lunar meteors. After 170 hours of searching the dark portion of the Moon with telescopes, Haas and his associates had detected 12 bright moving specks which began and ended at points on the Moon. During the same observations, four or five Earth meteors crossed the field of view. One or two of the lunar flashes may have been faint Earthbound meteors, but the laws of probability indicated that the rest occurred on the Moon.

The reason meteors seemed to be stopped more effectively in passing through the Moon's atmosphere than the Earth's will now be given. Measurements made during the Apollo missions indicated that a bulge exists on the far side of the Moon. This implies that the density and depth of the atmosphere on the near side are much greater than the average density and depth. It is significant that the near side is primarily comprised of the so-called maria. They were originally given this name because they have all the appearances of dried-up or drained-off oceans and seas. The far side was determined to be mostly mountainous, giving the Moon extremes of elevation greater than Earth's. This same condition would occur on Earth if the oceans and seas lost their water. If the average lunar atmospheric thickness is about the same as Earth's, the conclusion is that the atmospheric density on the near side of the Moon is greater than any place on Earth!

The startling implication of a dense atmosphere is that spacesuits and life-support systems might not be needed in most areas on the Moon if the atmospheric gases are the same as Earth's. This suggests that the Apollo astronauts may have been wearing extremely light backpacks since oxygen requirements could have been supplied by the Moon's atmosphere. It follows

Incredible Findings Concerning the Lunar Atmosphere

that the suits were probably only used during filming to propagate the cover-up. After completing the filming sequences, the astronauts could discard the suits and backpacks and go about their Moon exploration or other activities completely unencumbered. However, in other areas the gear may have been required, just as on the highest Earth mountains. If the Earth lost its oceans, many mountainous regions and high plateaus would no longer have sufficient atmosphere to sustain life. The atmosphere would seek the lowest level and fill up the ocean beds which reach depths of many miles. Since the Earth's oceans cover the majority of its surface, millions of square miles would probably become uninhabitable.

It is conceivable that life and vegetation could exist in certain regions despite the long lunar days and nights. Sheltered canyons and valleys at the right elevations and latitudes would not experience the extremes of temperature found in the uninhabitable areas. The extremely long days and nights occurring in the Earth's polar regions might produce very similar conditions to certain lunar regions, and significantly, life forms have adapted well to these extreme conditions on Earth.

The previous references to drifting clouds and mists suggest surface water. Observations indicate that cloud formations are more extensive in mountainous regions and inside craters where the moisture is trapped between natural barriers. Drifting clouds require winds to move them. In a vacuum, the discharged gases would diffuse out rapidly and would not drift.

Another indication of a dense lunar atmosphere was provided when Apollo spacecraft and lunar probes orbited the planet at an average distance of 70 miles above the surface. No specific reasons were given by NASA for choosing this height. In fact, if the Moon had no atmosphere, the best altitude for the Lunar Orbiter satellites would have been much lower. The Lunar Orbiters were sent to the Moon to take pictures; therefore, lower altitudes would have produced more refined maps of the surface. Even the Apollo Command Module orbited at this altitude. An atmosphere forces the minimum permissible orbit to an altitude where frictional air resistance does not slow down a vehicle or satellite substantially. The effects of an atmosphere on a low altitude spacecraft would be a quick decay in the orbit, causing the spacecraft to slow down, burn up, and crash. This is why Earth spacecraft, such as Skylab and other satellites, stayed more than 100 miles above the Earth. It seems that the orbital altitude chosen by NASA was probably due to the Moon's atmosphere since it prevented them from safely orbiting at any lower altitude for any significant length of time. This implies that the density of the Moon's atmosphere may be similar to Earth's.

One of the significant discoveries of the lunar program was that the Moon has a very feeble magnetic field. The existence of a lunar magnetic field did not clash with orthodox beliefs about the origin of a planet's magnetism because a small iron core could always be used to explain it and the size of the core could be adjusted to fit the extent of the magnetism found.

The most probable cause of a planet's magnetic field seems to be the rotation of charges which are present in its atmosphere and on its surface. These charges rotate with the planet; therefore, the intensity of the magnetic field generated would be directly proportional to the planet's rotational velocity. Since the Moon's rotational velocity is less than one percent of Earth's, it also follows that the Moon's field is less than one percent of Earth's. The Lunar Analysis Planning Team came to the consensus that natural remanent magnetism found in Moon rocks suggested that the Moon had a magnetic field strength at one time equal to several percent of Earth's.[13] They were still uncertain as to how it was generated.

A rotating planet can be compared to an electrical solenoid, which is a coil of wire, as shown in Figure 3. When a current is sent through the coil, a magnetic force is generated at right angles to the direction of the wire. Planets carry charges with them in their atmosphere and surface and this generates electrical currents in the direction of rotation, or east-west. The magnetic field is generated at right angles to this or in the north-south direction. Even though the charges are not necessarily moving east-west relative to the surface which is traveling with them, the magnetic field is still created because the planet itself is rotating. Other factors such as surface material, surface anomalies, caverns, winds, and so on would affect the direction and magnitude of local magnetic fields.

The explanation just presented explains many known facts regarding geomagnetism. For example, sunspots affect geomagnetism by altering the number of electrical charges in the atmosphere and surface. In addition, geomagnetism follows a 24-hour cycle due to the variable number of charged particles reaching Earth from the Sun. Evidence will be presented in the next chapter that the Moon has no iron core. Without an iron core, orthodox physicists would have difficulties in explaining lunar magnetism. However, the new approach is logical

Figure 3

A NEW THEORY OF PLANETARY MAGNETISM

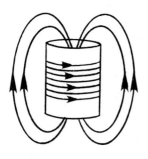

Planetary magnetism is generated in the same way as the field around a solenoid (coil of wire carrying a current). In each case, the direction of the magnetic field is at right angles to the direction of movement of charged electrons. A compass will align itself in the direction of the field. The solenoid shown here has electrons moving left to right; therefore, the magnetic field is in the up-down direction.

Electrons flow left to right through wires of the solenoid coil

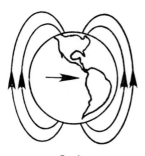

North

South

The magnetic field of a planet is at right angles to the direction of rotation of the charged particles carried along with it in the atmosphere and surface. Even though the particles are not necessarily moving with respect to the surface, the magnetic field is still generated in the north-south direction. Most of the particles probably come from the Sun.

All that is needed to generate a magnetic field is the movement of electrons or charged particles. This requirement is met in both a rotating planet and solenoid.

Direction of rotation of the Earth is shown by arrow.

Incredible Findings Concerning the Lunar Atmosphere 107

and completely adequate to explain planetary magnetism without iron cores.

Additional evidence of the Moon's substantial atmosphere is provided by statements which some astronauts made to the effect that stars are not visible above the atmosphere. This event was not widely discussed in keeping with the policies of NASA concerning events which deviate from standard beliefs. If this is true, then the lunar atmosphere would affect star visibility in a similar way to the Earth's atmosphere. Stars would then be visible only when viewing them through an atmosphere. On the Apollo 11 mission, less than 13,000 miles before reaching the Moon, Armstrong stated:

I see the crater Tycho fairly clearly. I can see the sky all around the moon, even on the rim of it, where there's no earthshine or sunshine.[14]

Collins then stated:

Now we're able to see stars again and recognize constellations for the first time on the trip.... The sky's full of stars.... it looks like it's night side on Earth.[15]

It was also stated that the Moon blotted out some of the Sun, enabling them to see the heavens more clearly.

The conditions to be expected in a vacuum were presented in the last chapter. References were cited which claimed that the stars would be visible day and night on the Moon without an atmosphere. In actuality, most stars may not be visible to the naked eye without the aid of an atmosphere. An atmosphere acts like a giant lens and causes starlight to spread out. Because stars are considerable distances away, giant telescopes can only resolve or detect the disks of a small number of them. Only a blur of most stars is seen through a telescope and this blur is largely due to scattering and bending effects of starlight passing through the atmosphere. So the human eye, being less effective, might only be able to detect a few of the brightest stars above the atmosphere. Therefore, the ones the astronauts could see with their naked eyes would only appear as tiny pinpoints of light as opposed to blurred or twinkling stars.

As a consequence of the above analysis, Collins' remark about their ability to see stars again may have had nothing to do with the Moon blotting out part of the Sun. Why couldn't the astronauts have looked in a direction away from the Sun, to an area of space not occupied by either the Earth or Moon? A significant amount of light diffusion should not occur in the space between the Earth and Moon. Therefore, if the astronauts were facing away from bright planets and suns, they should have been able to see stars if it were possible to do so without an atmosphere. When the astronauts reached the vicinity of the Moon, they would be able to see stars and constellations again through the Moon's atmosphere.

One of the most convincing photos showing a dense lunar atmosphere appeared in the 1971 *Encyclopedia of Discovery and Exploration,* on page 131 of the 17th Volume, *The Moon and Beyond* by Fred Appel. This picture, taken by the Lunar Module of the Apollo 10 Command Module in orbit above the Moon, is shown in Photo 11. The Moon's atmosphere appears as a dense band of blue on the lunar horizon. It closely resembles pictures of the Earth taken from Earth-orbiting satellites and spacecraft. This same band of atmosphere is what Neil Armstrong may have referred to in the previous quotation. He could not have been more precise in what he described, "I can see the sky all around the moon, even on the rim of it, where there's no earthshine or sunshine."

Incredible Findings Concerning the Lunar Atmosphere

It is appropriate to mention the difficulties the author encountered trying to obtain the photograph just referenced. A number of letters were sent to NASA in 1979 requesting the NASA photo number for this picture so that it could be ordered. However, no reply was ever received, even when photocopies were sent to NASA so that they would have no problem identifying it. NASA is supposed to provide this service free of charge to the public. Since this photo appeared elsewhere, it had already been issued. The author then requested the aid of the publisher of *The Moon and Beyond,* Aldus Books Ltd. in London, England, to obtain the NASA number.

In December 1979, David Paramor of Aldus Books answered the author's request and identified the NASA number as 69-HC-431. He indicated that NASA had given Aldus Books permission to use it in *The Moon and Beyond.* The author then ordered photo 69-HC-431 from NASA's photo contractor and sent a photocopy of the picture with the order to assure that the correct photo would be returned. When the order arrived, the author received a photo consisting of 24 separately labeled Apollo 4 rocket pictures on one transparency. The pictures were numbered starting with either 66 or 67, the year the Apollo 4 pictures were taken. However, the author ordered an Apollo 10 picture, 69-HC-431. It is also strange that a label with the number, 69-HC-431 had been placed on one corner of the transparency with masking tape, yet 69-HC-431 could not have possibly been the correct number. The first two digits of the NASA photo number are coded to use the last two digits of the year the photo was taken; hence 69-HC-431 was taken in 1969 and could not have been a photo of the Apollo 4 rocket. It is also significant that this was the only one labeled with masking tape out of over 50 photos ordered from NASA's photo contractor.

In July 1981, more than a year after the first attempt, the author again requested NASA's help in obtaining the photograph; a photocopy of it was sent; and even the number was sent, 69-HC-431. The author informed NASA of the problems encountered with the photo contractor; that Apollo 4 pictures were sent instead of the Apollo 10 photo. It is regrettable that as of the date this book goes to the printer, NASA has not responded in over a year-and-a-half to the author's requests. However, on the bright side, the author's experiences in trying to obtain it provided almost as much evidence for the NASA-military cover-up as the picture itself.

Before tackling lunar geology and the structure of the Earth-Moon system, another incident will be presented which demonstrates how far NASA was willing to go to suppress the facts. On the Apollo 15 mission, a hammer and a feather were supposedly shown to fall at the same rate. The objective was to demonstrate the vacuum condition on the Moon and Galileo's famous experiment on gravitational attraction, that objects of different sizes and weights fall with equal velocity. In view of all the evidence demonstrating a substantial lunar atmosphere, there seems to be only one reasonable conclusion: If the two objects hit the ground at the same time, the feather probably concealed a rather heavy object.

At the time of the Apollo missions, a number of observers noticed some of the discrepancies mentioned in this book. People like Bill Kaysing were convinced that the Apollo missions were faked and that all the telecasts and photos were made in some remote place on Earth. Kaysing wrote a book in 1976 entitled *We Never Went to the Moon*. If he still believes that the Moon is

Incredible Findings Concerning the Lunar Atmosphere

airless and has only one-sixth of Earth's gravity, then he is justified in concluding that the missions were faked. It does not seem probable that NASA, government officialdom, or the astronauts will deny his assertions. They will simply ignore them rather than face the impossible task of explaining away the contradictions.

By now, many readers may wonder why so much evidence of a dense Moon atmosphere would be ignored by the world's scientists for over a century. The reason is that a substantial lunar atmosphere means a high lunar gravity. In turn, a high lunar gravity means that the superstructure of conventional physics has a shaky foundation and might be seriously damaged if this fact is recognized. The military knows that these facts are the keys to understanding gravity and how to control it. In each case, vested interests are probably watching out for their own well-being at the expense of the general public.

CHAPTER 8

MOON GEOLOGY AND THE EARTH-MOON SYSTEM

Moon geology provides additional evidence that the Moon has a high surface gravity and a substantial Earth-like atmosphere. Other geological information collected by Apollo provides clues about the Moon's internal structure. When most people think about the Moon, they envision craters. However, craters are only one aspect of the Moon. Others deserving attention are the maria or seas, mountains, valleys, cracks, and river-like channels or rilles which meander across the surface and through lunar valleys. The craters often obscure and obliterate the details just mentioned. It is as though the Moon had an Earth-like surface at one time, but was subsequently bombarded by tremendous forces and essentially destroyed.

Most conventional scientists believe that the Moon could never have had an atmosphere because of the weak one-sixth gravity. They then rationalize that all lunar features are the result of meteors, volcanoes, or solar wind bombardment. Without an atmosphere, there would be no rivers, weathering, and so forth. They claim that the Moon is dead and always has been. The majority of orthodox scientists will not even consider the possibility that civilizations have existed that could destroy a planet's surface or reduce an entire planet to rubble with sophisticated weapons. This attitude then forces them to find natural causes for everything observed on the Moon, Earth, and other planets. This has been the trend of scientific interpretation of lunar geological discoveries to the present. It should also be noted that if civilizations of the past had the capabilities just mentioned, some of them are still undoubtedly in existence today. With a highly developed space technology, the obliteration of one or two planets would not destroy life on all planets. Many survivors of the conflict would occupy other planets and space colonies. If conventional scientists considered intelligent intervention in the Moon's geology in the distant past, they would soon be speculating on remnants of the civilization and subjects like UFOs.

The primary debate among Moon geology experts has been between those who believe the Moon has a hot interior and those who believe that it has a cold interior. Others believe that at one time it was hot but has since cooled. Some believe that most of the craters were produced by meteors, while others think they are volcanic in origin. One group postulates that the maria or oceans were created by volcanic lava which seeped up from the interior to create these basins after the Moon was bombarded by meteors. These same theorists believe that the craters are not as deep as they should be from meteor impact due to the same seeping lava. Without consideration of the Moon's high gravity, atmosphere, and the possibility of intelligent intervention, conventional scientists are working with very limited information.

An important discovery which is essentially unknown to orthodox physicists and geologists is the subtle transmutation or transformation of elements into other elements without the release of radioactive by-products or radiation. The researcher Louis Kervran discovered the fundamental relationships which are responsible for the formation of mineral deposits in the Earth. He also determined that biological organisms constantly transmute certain elements into other elements in a manner unknown to most nuclear scientists. His findings cannot be refuted, yet his work has been ignored by the scientific world because the findings do not fit in with conventional theory. Geologists know that certain minerals are found in conjunction with other minerals in varying proportions. Kervran was able to show that deposits have varying proportions of these minerals because their constituent atoms actually change from one element to another over a period of time without emitting harmful radioactive particles or by-products. The implications of Kervran's findings radically change the fields of physics, geology, and many other scientific disciplines. Mineral deposits and soil can transmute many times faster than conventional science claims. Therefore, the radioactive methods used to date Moon rocks may not give results representative of the true history of the Moon. Similarly, dates given to Earth rocks may be incorrect. Kervran's book, *Biological Transmutations,* gives a summary of his incredible findings.

    The Earth's atmosphere is the primary factor responsible for surface erosion and weathering. However, our atmosphere would become very ineffective in causing erosion if the surface water were lost. The Moon's geological features indicate that the Moon had abundant surface water in the past because the lunar hills and mountains are mostly rounded and weathered, and many river-like watercourses and rilles or cracks appear on the surface. Photo 7 shows the rounded and weathered hills in the Hadley Delta-Apennine Mountain region as well as the illuminated atmosphere. Apollo 15 astronaut Scott is standing in the foreground. Photo 12 shows Scott standing next to the Rover on the edge of Hadley Rille (as much as 1,200 feet deep in places) which could easily be mistaken for a water-formed dry gulch or river canyon on Earth. Photo 13 was not taken on the Moon but in Eastern Washington State by the author. It shows the same kind of weathering as many Moon photos. Photo 14 looks North across the Apennine Mountains with a segment of Hadley Rille winding across the bottom center. Three-mile high Mount Hadley is the shadow-sided peak above and to the right of the "chicken's head" shape of Hadley Rille. Apollo 15 landed on the plain just above the "chicken's beak" bend in the Rille. Craters Autolycus (25 miles wide) and Aristillus (36 miles wide) appear in the upper left near the eastern edge of the Sea of Rains (Mare Imbrium). These Apollo 15 photos provide convincing evidence that the Moon has an Earth-like gravity and atmosphere to account for the geological formations and erosion. An Apollo 17 picture of astronaut Schmitt standing next to a huge split boulder is shown in Photo 15. The rounded hills surrounding Littrow Valley and the steep slopes of East Massif appear in the background about 5 miles away. Littrow Valley is on the edge of Mare Serenitatis. The same weathering evidence appears here as in the Hadley Delta-Apennine region which is on the opposite side of Mare Serenitatis.

  A photo of the Moon's Alpine Valley taken by the 1967 Lunar Orbiter IV probe is shown in Photo 16. Starting in the upper right section of the photo, the Alpine Valley runs for 90 miles before it stops at the northeastern edge of Mare Imbrium (bottom left). In the middle of the valley is what appears to be a dry riverbed flowing through it into a dried-up sea (Mare Imbrium). Scientists have referred to such meandering channels as sinuous rilles and some of them even believe that such rilles were carved by water in spite of the Moon's alleged low gravity and consequent vacuum condition. The above photos point out the problems orthodox scientists have in explaining the Moon's features. If there is no atmosphere, how could there ever have been water, clouds, and rivers? There is only one reasonable conclusion from the evidence: The Moon had an atmosphere and abundant water on the surface. Therefore, the Moon had a high gravity to hold the atmosphere. But if it had a high gravity at one time, it must still have one. Since it still has a high gravity, it still has a dense atmosphere.

Is there any other evidence that water once existed on the Moon? An article entitled "What Is It Like to Walk on the Moon?" by Apollo 15 astronaut David Scott appeared in the September 1973 *National Geographic*. In it he stated:

A dark line like a bathtub ring smudges the bases of the mountains.[1]

These lines have been referred to as watermarks because they look like those seen along a shore. However, scientists are puzzled because water supposedly cannot exist on the Moon in the first place. The astronauts observed that Mt. Hadley displayed distinct linear markings which inclined toward the northeast at a 45-degree angle.[2] If these same lines had been found on Earth, they would have been interpreted as sedimentary deposits. However, according to conventional scientists, the Moon has no known processes to account for their existence. On the Apollo 16 mission, the astronauts discovered that Stone Mountain appeared terraced and generated the same kind of lines seen by Scott and Irwin on the Apennines.[3]

What happened to the water? Some of the similarities between the Earth and the Moon were pointed out in the chapter on atmosphere. The Moon has vast areas on the side facing the Earth which appear to be like oceans. It is significant that the Moon's oceans have the lowest elevations in relation to the rest of the Moon. Apollo 15 determined that the side facing the Earth is between three and six kilometers lower than the opposite side. In addition, the opposite side is primarily mountainous with few maria or oceans.

Since the Moon probably has a high gravity and a dense atmosphere, the water could not have easily escaped into outer space. It couldn't have gone into the atmosphere because there is no substantial quantity of water there now. The one remaining place for it to have gone is into the Moon's crust. However, the only way this could have happened is if the Moon's crust is cavernous in nature. In order for the water to have entered underground caverns, openings and cracks must have been created first. This might have happened if the Moon were bombarded by large enough meteors or sophisticated weapons. Once the initial cracks or rilles appeared on the ocean bottoms, the oceans would literally drain into the crust, leaving vast watercourses, dry riverbeds, and eroded rilles all over the Moon. The dry ocean basins would then take on the characteristics of Death Valley.

If the Moon has a cavernous structure, what evidence exists to demonstrate this? Interestingly enough, Moon probe experiments discovered mascons and Apollo missions conducted seismic experiments. Mascon is short for mass concentration. These were hypothesized by scientists to explain a perplexing discovery: Moon probes were pulled from side to side and moved up and down as they passed over certain portions of the Moon. Scientists deduced that large meteors were buried a short distance under the surface which caused increases in the local gravitational fields. Some of these meteors were postulated to be 402 miles in diameter and 2½ miles thick to account for the gravity variations. Why are pancake-like meteors floating through space? The higher density of the buried meteors supposedly increased the local gravity. The variations were supposedly higher than those found on Earth.

The explanation for mascons presented by scientists creates many contradictions. First, the gravity increases are found over the lunar oceans and seas. These are the smooth areas of lowest elevation with essentially no craters. Meteors of the sizes propounded should have produced devastating craters in the oceans and seas. This was explained away by assuming that molten material came to the surface and filled in the holes. A second problem is that if the Moon's upper crust was so molten, the meteors should have sunk much deeper into the Moon instead of being stopped at the surface. Some scientists contended that mascon unevenness could not exist in a hot body. In addition to the above problems, the Moon has a lot of basalt substance produced by volcanism. The meteoritic mascons and a hot Moon, indicative of volcanism, do not agree.

The above problems seriously undermine mascon theory. This implies that other explanations for the gravity variances must exist. Two factors will now be considered. First, the gravitational attraction over large bodies of water on Earth has been found to be greater than over land masses.[4] Second, the assumption of a cavernous Moon can be used to explain the gravity variations, but orthodox scientists ignored this explanation, at least in the official reports. If the Moon's oceans drained into the crust and partially filled up vast caverns, an explanation for the gravity variances is generated. This could also account for some of the watercourse rilles and missing surface water at the same time.

The meteoritic versus volcanic crater debate is another question which is difficult to resolve without consideration of possible intelligent intervention and the Moon's cavernous nature. Ranger 7 gave scientists their first headache by indicating a surface which looks like a rolling desert instead of the jagged terrain expected on a planet without an atmosphere. Surveyor 1 landed in Oceanus Procellarum and the pictures indicated lunar soil similar to terrestrial soil without water. Surveyor 5 landed on Mare Tranquillitatis near the top of a crater. The chemical analyzer indicated that the soil was a type of basalt which forms the rocks of ocean floors and ridges on Earth. The findings of Surveyor 5 also indicated that the amount of magnetic material was insufficient if the surface were meteoritic.[5]

Surveyor 6 produced results similar to Surveyor 5 and led scientists to believe that the chemical composition was common to the lunar maria. Surveyor 7 landed in the highlands and the analysis indicated a less dense material than the mare basalt. In addition, all of the Surveyors indicated that the most abundant lunar elements were oxygen and silicon, as on Earth. Therefore, the Moon was determined to be an evolved planet, not just a source of meteorites.[6] Even so, the identification of basalt in the mare basins, indicative of volcanism, did not convince the impact theorists that they were wrong. Consequently, the impact versus vol-

Moon Geology and the Earth-Moon System 121

canic theory was still unresolved among scientists at the end of Surveyor.

Neil Armstrong consistently discovered glassy patches in the bottoms of small craters. Astronomer Thomas Gold theorized that the Moon had been scorched by a flare-up of the Sun.[7] An enigma was created because the patches were apparently untouched by the constant bombardment of micrometeorites and solar particles. In addition, the glaze rested on tiny pinnacles and stems of the lunar soil. Gold surmised that the flare-up had therefore occurred less than 30,000 years ago and lasted for only 10 to 100 seconds. An alternative explanation is that sophisticated weapons were used to bombard the Moon less than 30,000 years ago. The undisturbed, glassy patches would indicate that micrometeorites are not reaching the surface; therefore, they must be stopped in passing through a dense atmosphere. This is additional evidence debunking the meteorite theory of lunar erosion and supporting the other evidence of the Moon's Earth-like atmosphere and gravity. The location of the Apollo 11 landing was a lowlying mare basin near the equator, Mare Tranquillitatis. This would be an area with minimal lunar weather, such as in the Southwestern United States where the military has stored aircraft for years with few deteriorating effects on the planes. In these areas on the Moon, items would be undisturbed for perhaps aeons. The relatively still atmosphere would not disturb the terrain and most meteorites could not penetrate through the atmosphere.

In *Strange World of the Moon,* Firsoff suggested that the volcanic theory of lunar craters is difficult to substantiate because there is not much evidence of fire and smoke, or of ash and lava.[8] Since the Moon's surface seems to resemble the Earth's in so many ways, it is reasonable to conclude that its period of extensive volcanic activity may be nearly over. The rounded surface features and weathered appearance, as well as the extensive oceans or maria, indicate that volcanism has not been a significant factor on the Moon since the formation of those features. Extensive volcanism would have disturbed the smooth maria and created a more rugged appearance. Since many of the recognizable, Earth-like features have been obliterated by overlying craters, the craters themselves were not created in the period of extensive volcanic upheaval. They had to come after the surface conditions stabilized.

In a book by Don Wilson entitled *Secrets of Our Spaceship Moon,* the writer referred to the NASA publication entitled *Apollo 17: Preliminary Science Report* (1973).[9] In it, the conclusion derived from the Apollo 17 mission was that sufficient evidence had been accumulated to indicate that volcanic activity in the last three billion years was virtually nonexistent or highly restricted. This provides additional evidence that frequent lights seen on the Moon are not due to volcanic outgassing.

The above analysis indicates that many of the Moon's extensive craters may not have been caused by either volcanoes or meteors. Most of these craters were evidently created after the surface reached a state of maturity which made it Earth-like in appearance. The Moon must have had extensive weather, oceans filled with water, and so on. A look at the probable history and origin of the Moon is reserved for a later chapter.

Seismic experiments were conducted to determine the nature of the Moon's crust and internal structure. Sensitive seismometers were left on the surface by lunar probes and Apollo missions. When the Lunar Module and other objects were sent crashing into the Moon, shock waves were recorded and experts could interpret the data. The results of the experiments were not anticipated by scientists. The Apollo 11 seismometer indicated that the Moon was relatively quiet. To some scientists, this meant that the Moon had a small rather than a large molten iron core. Others believed that the Moon did not have a core at all. Apollo 12 carried a much more sensitive instrument that was designed to operate continuously. After the discarded LM hit the Moon about 40 miles from the landing site, the three long-period seismometers picked up a sequence of reverberations which lasted for more than half-an-hour. This meant that the structure of the Moon was quite rigid because it resonated like a bell when struck. Some scientists contended that the Moon is solid without any liquid inside. This was another blow to the molten core hypothesis. Others concluded that the Moon was hollow, but this did not seem to agree with gravity theory.

The Moon's high surface gravity indicates that there must be a serious flaw in Newton's Law of Universal Gravitation. This flaw is a key to understanding the true nature of gravity for the first time. Since Newton first formulated his Law in 1666, not a single explanation for the nature of gravity has been generated and accepted by the academic community. Even Newton did not claim to understand the nature of gravity. He merely attempted to describe the effects of gravity on falling bodies in mathematical terms. Newton assumed that whatever this mysterious gravitational force was, it somehow acted on all matter uniformly and was not scattered or attenuated as it penetrated the matter of planets thousands of miles below the surface. His theory implied that gravity was something associated with every particle of matter, regardless of its position in space. The probable major flaw in his Universal Law is the inherent assumption that gravity effects penetrate matter with no interacting, scattering, or multiplying effects other than the normal decrease due to the inverse-square law. Without these effects, a gravitating body's attraction for any other body is not affected by matter placed in between these bodies; hence, gravity could not exert any forces on masses. Since gravity does exert forces on matter, these effects must exist. Therefore, Newton's Law of Universal Gravitation violates the law of conservation of energy. Since forces require energy, the insertion of matter between two gravitating bodies will generate energy interactions and cause the gravitational forces between the two outer bodies to decrease, unless there are additional gravitational effects generated by the inserted matter.

The scattering of gravity seems to be demonstrated when mountain masses do not pull plumb bobs out of line to the extent demanded by Newtonian Gravitation theory. Geologists have tried to explain this by assuming that the average density of matter under mountains is less than that of matter under oceans. A more likely explanation is that gravity effects produced by matter inside mountains are partially dispersed or attenuated by overlying masses. This is one factor which may account for the differences. The above information and the Moon's high gravity suggests that gravity is produced by a very penetrating radiation. Although it penetrates matter to considerable depths, the ability is still limited.

It is significant that the Moon's high surface gravity implies an impossible mass for the Moon if the Universal Law is used. If 64 percent of Earth's gravity is

Moon Geology and the Earth-Moon System 125

assumed, the Moon would need an average density of 13.0 gm/cm³ (grams per cubic centimeter) according to the Gravitation Law. This is greater than the density of lead which is nearly 50 percent heavier than iron.

From a determination of the center of mass between the Earth and Moon, the Earth would require 81.56 times the Moon's mass which would give it a density of 21.5 gm/cm³. This is about twice the density of lead; therefore, even the iron core hypothesis cannot solve the mass enigma. It should be noted that the flaw in Newton's Law of Universal Gravitation inspired the hypothesis of the Earth's iron core. Once the Earth's mass had been assumed, the Moon's mass was determined by the gravitational force at its surface. It is probable that only a limited thickness of the Earth's crust contributes to the majority of the Earth's surface gravity due to the scattering of gravity radiation originating from masses below a certain depth. This implies that the Earth's mass cannot be accurately predicted using the conventional method. If the planets had empty or hollow centers, the surface gravity might not be much different than if they had iron cores, or even lead cores. This seems to explain why the Moon has such a high gravity for its size. Because of this effect, the conclusion is that Newton's Universal Law overstated the mass of the Earth in the first place.

The Earth's iron core was hypothesized because the average density of the crust was not adequate to account for the whole Earth's predicted mass. The Earth's crust has an average density of 2.7 gm/cm³ compared to the Moon's average crust density of 3.3 gm/cm³. To satisfy Newton's Gravitation Law, the Earth's average density had to be 5.5 gm/cm³. This led to the Moon's average density of 3.34 gm/cm³. The slight difference between the Moon's surface density and the predicted total average density enabled a small iron core to be postulated to account for the slight magnetism found. This is where orthodox Moon theory stands today. In the last chapter, geomagnetism was explained using a theory which does not depend on the existence of an iron core. Scientists believe that they have verified the existence of the Earth's iron core by measuring shock waves reflected off it and through it. However, they might also get a reflection off large caverns inside the Earth or off the inner surface if the Earth were a hollow shell. Similar experiments were conducted on the Moon and shock wave measurements convinced scientists that they had found the Moon's mantle. However, the evidence provided by Apollo seismic experiments also points to the conclusion that the Moon is hollow and relatively rigid.

It is not commonly known that the Earth displays the same bell-like ringing or reverberation as the Moon. Since the Earth is 81.56 times more massive than the Moon, it takes a much larger explosion or shock wave to generate this effect. Joseph Goodavage referenced such occurrences in his book, *Astrology: The Space Age Science*.[10] He mentioned that the ringing effect was recorded during the May 22, 1960 Chilean earthquake. This was supposedly the most violent earthquake that had been recorded since the establishment of official world records in 1881. Goodavage provided a description of the effect which was given at the 1961 World Earthquake Conference, held at Helsinki, Finland. The description stated that the shock was so severe that the entire planet rang like a bell. The ringing continued for a considerable length of time in a regular series of slow impulses which were recorded at various independent seismic stations. Goodavage also noted that the planet rang again as a result of the Anchorage, Alaska earthquake of March 27, 1964. It seems somewhat hard to believe that scientists were so appalled in finding that the Moon rang like a bell. After all, the Earth displays the same characteristic.

The hollow Moon concept was extensively explored by Don Wilson in the previously mentioned book, *Secrets of Our Spaceship Moon*. In it, Wilson referenced a *Saga* interview conducted by Joseph Goodavage with Dr. Farouk El Baz, a former NASA geologist who trained the astronauts.[11] According to El Baz in the *Saga* article, not all NASA discoveries were announced: For example, he mentioned that there were many undiscovered caverns within the Moon; that several experiments were performed to see if caverns existed along with subsurface ice.[12]

What does the hollow Moon concept do to the Newtonian concept of gravity? It indicates a lower Moon mass than that predicted by the famous Newtonian Law. It provides evidence for the limited penetrating ability of gravity as previously explained. Finally, a hollow Moon also implies that the Earth is hollow. Before providing additional evidence, an important point needs to be stressed once again: Scientists have determined the center of mass of the Earth-Moon system. This enables them to accurately calculate the ratio of Earth mass to Moon mass. The problem then is to correctly determine the mass of either one of the planets (Earth or Moon). But if they are hollow, the shell thickness must be known to determine the planet's volume. In addition, the average density and extent of caverns must be known. Without iron cores to worry about, the average density can be approximated based on the density of the crust, but the shell thickness is difficult to estimate, even with a number of incredible Earth photos. One of these is presented for the reader's inspection in the next chapter.

Photo 1   Bean standing next to Surveyor 3 with the Lunar Module in the background on the Apollo 12 Mission. (NASA Photo)

Photo 2   Young jumping up from the lunar surface on the Apollo 16 Mission (NASA Photo)

Photo 3 Bean carrying the ALSEP on the Apollo 12 Mission (NASA Photo 69-HC-1341)

Photo 4   Irwin with the Lunar Rover and Mt. Hadley in the background on the Apollo 15 Mission (NASA Photo 71-HC-1140)

Photo 5 Aldrin standing next to the deployed Solar Wind Composition Experiment on the Apollo 11 Mission (NASA Photo)

Photo 6   The lunar surface with the Lunar Module and tire tracks on the Apollo 14 Mission (NASA Photo 71-HC-277)

Photo 7 Scott standing on the slope of Hadley Delta with the Apennine Mountains in the background on the Apollo 15 Mission (NASA Photo 71-H-1425)

Photo 8 Duke collecting samples on the rim of a crater with the Rover in the background on the Apollo 16 Mission (NASA Photo)

Photo 9 A halo around Bean on the lunar surface on the Apollo 12 Mission (NASA Photo 69-HC-1347)

Photo 10 A frame taken from the movie sequence of Mitchell's descent down the Lunar Module ladder on the Apollo 14 Mission (NASA Photo)

Photo 11 A picture of the Apollo 10 Command Module taken by the Lunar Module with the Moon's horizon in the background (NASA Photo)

Photo 12 Scott standing next to the Rover on the edge of Hadley Rille on the Apollo 15 Mission (NASA Photo 71-H-1426)

Photo 13 A picture taken by the author showing eroded hills in Eastern Washington State

Photo 14   A view of the eastern edge of the Sea of Rains with Hadley Rille and the Apennine Mountains taken on the Apollo 15 Mission (NASA Photo)

Photo 15 Schmitt standing next to a split boulder with rounded hills in the background on the Apollo 17 Mission (NASA Photo)

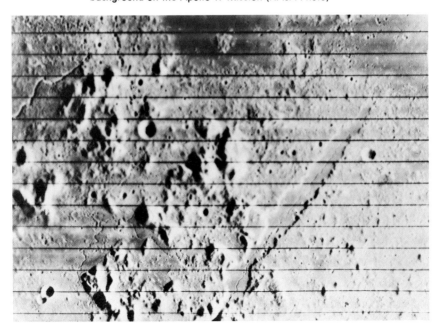

Photo 16 The Moon's Alpine Valley photographed in 1967 by the Lunar Orbiter IV probe (NASA Photo 67-H-897)

Photo 17　An Applications Technology Satellite Photo of the Earth taken 22,300 miles above Brazil in 1967 (NASA Photo 67-HC-723)

Photo 18　A reproduction of the color TV transmission of the Lunar Module lift-off from the lunar surface on the Apollo 17 Mission (NASA Photo 72-HC-903)

Photo 19   A reproduction of the color TV transmission of the Lunar Module lift-off from the lunar surface on the Apollo 16 Mission (NASA Photo 72-HC-274)

Photo 20   A reproduction of the color TV transmission of the Lunar Module lift-off from the lunar surface on the Apollo 16 Mission taken moments after Photo 19 (NASA Photo 72-HC-273)

CHAPTER 9

SATELLITE PHOTOS AND BIZARRE EARTH FINDINGS

Incredible as it may seem, a number of satellite photos of the Earth released to the public as early as 1967 show evidence of what appears to be a deep depression into the Earth in the North Polar region. A depression cannot be put into a sphere without causing a flat spot to show up in its outline when viewed from a certain angle. If there is such a depression, a photograph from a satellite in the right position would show the Earth to be out of round by a surprising amount. Other photographs would provide a three-dimensional perspective of the outline.

A photograph of the Earth taken by the DODGE (Department of Defense Gravity Experiment) satellite, 18,100 miles above the equator, appeared in the November 10, 1967 issue of *Life* magazine.[1] An approximate flat spot in the Earth's outline (not caused by photo cropping or the Sun's angle), about 1,600 miles across, is clearly evident in the North Polar region. It looks as if a large part of the Earth had been sliced off and thrown away. The author was not able to obtain permission to reproduce this photo; however, it is available for inspection at most libraries.

The DODGE photo is interesting, but a view at a better angle is needed to show more details. Such a picture, taken in 1967 by the Applications Technology Satellite III from its stationary position over the equator above

Brazil, is shown in Photo 17. It was taken from 22,300 miles out and shows what appears to be the outline of a large depression or hole in the North Polar region. The greater distance and the angle of the satellite's orbit would necessarily provide the three-dimensional effect seen if a depression or hole exists. Instead of having an abrupt edge, it appears to taper in gradually.

Critics will attempt to discredit these photos, insisting that the appearance of the depression is an illusion resulting from photo cropping, cloud formations, the Sun's angle, the North Polar ice sheet, open channels between the ice, etc. The author wishes to emphasize that a great deal of additional evidence will be presented in this chapter which does not depend on photographs. Therefore, the photographic evidence is only supportive of the remaining information. It is important for the reader to keep in mind that the Earth, or any other planet, may still be hollow without polar depressions. The information and evidence to be presented here is so incredible that the author does not expect the reader to easily accept the concept. A great deal more evidence needs to be produced before this idea can be readily accepted. The limited information presented here should be read with an open, inquiring mind in the interest of learning more about the subject and so as to be able to evaluate new evidence if it surfaces in the future.

In view of the NASA cover-up, it is not surprising that photographs of the Earth released to the public after 1967 (that the author has seen) do not show evidence of this entrance or depression, nor does it seem to be a coincidence that satellite pictures taken over the poles are not readily available to the public. The author made inquiries to NASA organizations and could not obtain any such polar photographs. One response to the

Satellite Photos and Bizarre Earth Findings 131

author's request from the NASA Technology Application Center was the following, "We do not have any satellite images taken over the poles." Significantly, this branch of NASA did not offer any help or suggestions on where to obtain them. It is common knowledge that satellites exist which are in orbits near the poles, and photographs should be readily available. There should be no reason to restrict or classify their distribution if NASA and the military have nothing to hide from the public.

It is important to stress that the DODGE photo and the ATS III photo were taken from slightly different angles, at different distances above the Earth, and at different times. Since the first photo mentioned shows an approximate flat spot about 1,600 miles across and the other photo indicates that the depression itself may be about 800 miles in diameter, the thickness of the Earth's shell can be very roughly estimated with the assumption that it is a hole. It is assumed that a cross section of the hollow Earth has the appearance depicted in Figure 4. Since the depression or entrance shown in the photographs appears to taper in gradually, it is assumed that it forms an approximate semicircle with a radius of around 400 miles. The thickness of the Earth's crust would then be approximately 800 miles and the area on the inside of the Earth would be over 63 percent of the outer surface area!

The gravitational pull on the inner surface would be in the direction of the inner surface due to the limited penetrating ability of gravity radiation. Furthermore, the force on the inner surface would probably be nearly the same as on the outer surface. A plane or ship traveling into the Earth's interior through the North Polar entrance would not feel any significant change in the pull

Figure 4

A CROSS SECTION OF THE EARTH

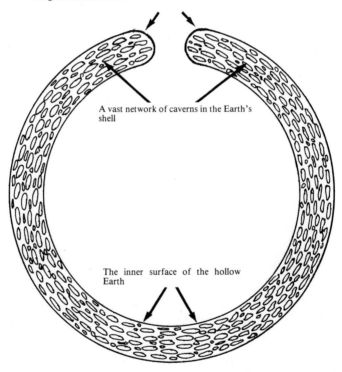

Satellite Photos and Bizarre Earth Findings   133

of gravity in passing from the outer to the inner surface. However, the curvature of the Earth would seem much greater. It follows that at a certain distance between the Earth's outer and inner surfaces gravity would be nullified. Objects or matter in this region of the Earth's shell would be weightless with no greater tendency to fall toward the outer than the inner surface. From a practical standpoint, a planet designed like this represents a feat of engineering genius. It produces a much larger surface area without using nearly as much mass as a solid planet would. In fact, the mass of the hollow Earth, assuming that the entire Earth has the density of its crust, would be less than one-fourth of the mass predicted by Newton's Law of Universal Gravitation.

The inferences about the Moon's mass are just as startling. Since the center of mass between the Earth and Moon is still reasonably accurate, the Earth's mass would be 81.56 times the Moon's. This implies that the Moon would also have less than one-fourth of the mass originally attributed to it. Assuming that its average density is the same as its crust, it would have a shell only 95 miles thick! This would cause it to ring like a bell with little provocation as the Apollo seismic experiments demonstrated. If the Earth and Moon have the same percentage of caverns in their crusts, the above thicknesses given for the two planets may be reasonable.

In the above calculation, it was assumed that the entire shell thickness of each body had the same density as its crust. It seems more reasonable that the Earth and Moon would have nearly the same overall density. If this is assumed, the corresponding lunar shell thickness would be 120 miles. In either case, if the Earth has a shell 800 miles thick, the Moon would have one seven to eight times thinner. Therefore, planets may be relatively fragile structures. This suggests the terrifying possibility that superweapons could reduce a planet to rubble! The evidence is practically overwhelming that Earth continents are drifting apart. At present, the theory of continental drift is widely accepted by orthodox science. However, without a hot, molten core, there seems to be no explanation as to why continents would drift apart. A new look at the phenomenon must consider the nature of gravity and the tremendous compressional and tensional forces generated in the Earth's shell which cause fractures and the sliding and folding of strata over each other. Continental drift might indicate that the Earth is slowly expanding over aeons of time. Extensive evidence for this contention is provided by the well-known scientist Pascual Jordan in *The Expanding Earth,* written in 1971. However, Jordan did not explain how a solid planet could expand to more than double its original diameter and still remain a solid sphere. If the Earth's original diameter doubled, its mass or volume would have to increase by a factor of eight. Since the Earth has only a limited amount of mass, it could not double in diameter without caverns or a hollow condition being generated to compensate for the growth in the volume. When the diameter doubles, the volume increases eight times; therefore, the present Earth would have a solid shell only 170 miles thick if it started out as a solid sphere 3,960 miles in diameter. It seems that a hollow, spherical Earth could not expand indefinitely without major depressions or openings being created. These would probably occur in the shell near the axis of rotation at approximate antipodes. This could explain why depressions or openings into the Earth's interior might exist near the poles.

Satellite Photos and Bizarre Earth Findings 135

There seems to be a simple explanation why the Earth may be expanding. As a simple example, when water in a glass is twirled around, it tends to be thrown outward from the center forming a void or vortex in the middle. The Earth is also made of material which will stretch and deform under similar circumstances. Objects on the equator spin at 1,000 miles per hour due to the Earth's daily rotation. Over aeons of time, the Earth might slowly stretch and expand due to this force. Consequently, it would form a hollow sphere, even if it started out as a solid ball. The gradual expansion would cause continents to drift away from each other.

Volcanoes have been explained by scientists using the hot, molten interior idea. However, it seems probable that the same forces and energies responsible for gravity and continental drift are involved in the generation of molten lava at pressure points in the Earth. A molten interior does not seem necessary to explain Earth phenomena such as volcanoes and continental drift.

Additional evidence for the existence of polar depressions or openings is provided by the Earth's magnetic field in these regions. Science has never explained why the magnetic poles are so far away from the geometric poles. The north magnetic pole is supposed to be located at about 76° north latitude, 100° west longitude; while the south magnetic pole is at about 66° south latitude, 139° east longitude, at the present time. Magnetic poles are thought of as points where the magnetic inclination, or the angle at which a compass needle points downward, is 90°. These magnetic poles would coincide with the geographic poles if orthodox beliefs about the shape and structure of the Earth were correct.

The existence of magnetic poles is probably misleading since they actually seem to be broad areas around a closed curve where the magnetic inclination is almost 90°. The south magnetic pole has supposedly shifted from 72° south latitude, 155° east longitude, where its location was first determined by Shackleton in 1909, to its present location at 66° south latitude, 139° east longitude. However, it is probable that there has been no radical shift in the magnetic pole at all in this time period. It seems more likely that the 90° inclination has been measured at a number of places coinciding with what could be areas near the southern depression.

Geomagnetic maps show lines of equal magnetic inclination centered around the supposed location of the magnetic poles. Significantly, an inspection of the previously shown ATS III Earth photo indicates that the location of the northern depression may correspond to the approximate location of the north magnetic pole shown on geomagnetic maps. It appears that the assigned location of the north magnetic pole may have been derived by averaging the various measurements made near the edge of this northern depression. It is also important to note that auroras are nearly concentric with the lines of equal magnetic inclination. The greatest auroral displays are not centered around the geographic poles but around the magnetic poles.

In the October 1947 issue of *National Geographic,* an article written by the late Admiral Byrd entitled "Our Navy Explores Antarctica" mentioned the warm winds coming from the vicinity of the Balleny Islands.[2] The Balleny Islands are near the curve of equal magnetic inclination where the inclination is close to 90°. From the location of the south magnetic pole, it follows that the southern depression would have to be located approximately 1,000 miles from the geographic pole. The warm winds Byrd spoke of may have been air currents coming

Satellite Photos and Bizarre Earth Findings 137

from this interior region, assuming it is that warm. Byrd also discussed a number of ice-free areas found with fresh water lakes amidst brown, barren rock hills. One area near the Queen Mary Coast of Wilkes Land was entirely free of ice over at least 300 square miles! Red, blue, and green algae were found in the lakes with warmer water than the Antarctic ocean. According to the article, no evidence was found of volcanic or hot spring activity; therefore, the warm areas were attributed to prevailing winds or solar heating during the summer months which kept the region free of snow and ice. The origin of the warm winds occurring in localized areas was not discussed.

Long before Admiral Byrd began his Arctic and Antarctic expeditions, explorers encountered baffling phenomena in their attempts to reach the North Pole. William Reed wrote a book in 1906 about their experiences entitled *The Phantom of the Poles*. It was found that many birds in the higher latitudes had a tendency to migrate north instead of south and that warm winds were often experienced by Arctic explorers. However, it seems that most of the later Arctic explorers had not read his or other books on the hollow Earth. These contentions about the Earth were not taken seriously because the ideas were in conflict with the foundations of basic physics.

Evidence that North Polar explorers were going into a sharply curving depression in their attempts to reach the pole is provided by their rate of travel. Frederick Cook claimed that he reached the pole on April 21, 1908 and Robert Peary claimed that he reached it on April 6, 1909. Each asserted that the other never reached it, but discrepancies surrounded both explorers. Cook had no legitimate witnesses and neither did Peary. Cook was doubted when he claimed to have made 15 miles a day, yet Peary allegedly made over 20 miles a day. After Peary neared the 88th parallel, he made 25 miles the first day, 20 on the second, 20 on the third, 25 on the fourth, and 40 miles on the fifth day. The conditions of travel were supposedly more difficult than those farther south where Peary averaged only 20 miles a day. In fact, even Peary made statements about the difficulties in maintaining walking speed with a dogsled due to adverse conditions.

It is unreasonable to believe that Peary could have traveled 270 miles from 87° 47' north to the pole and back in seven days. After he had been given credit for the discovery, a Congressional investigation later decided that his achievement was "not proven". The reason Peary and Cook seem to have covered such phenomenal distances is understandable if the curvature of the North Polar region is considered. Position checks with respect to the stars and Sun would tell them they had traveled much farther than they actually had if they still believed they were traveling on the surface with a radius of curvature of 4,000 miles. Other explorers noted that the horizon seemed to change more rapidly than they expected. For example, in *The Phantom of the Poles,* Reed quoted the explorer Greely concerning this phenomenon as follows:

> As we neared each projecting spur of the high headlands, our eagerness to see what was beyond became so intense at times as to be painful. Each point reached, and a new landscape in sight, we found our pleasure not unalloyed, for ever in advance was yet a point which cut off a portion of the horizon and caused a certain disappointment.[3]

Explorers also noted what they termed the "water-

Satellite Photos and Bizarre Earth Findings 139

sky" which was an exact reflection in the sky of what was ahead of them on the surface. They merely observed the sky and used it as a map to plan their course of travel. This strange condition could probably only occur if the Earth curved rather sharply; otherwise, the reflecting water vapor in the sky would be too far away to enable explorers to see the reflection.

Nansen was one of the early explorers who evidently traveled a considerable distance into the northern depression. At one point in his voyage, Nansen sailed north continuously for 15 days and should have gone over 1,200 miles past the pole. At that position, only one star was seen and for a period of time it remained straight overhead. Perhaps Nansen and his group were far enough inside to shut out the view of most of the other stars. In *Farthest North,* Nansen stated the following about it:

The sun had long since gone down behind the sea and the dreamy evening sky was yellow and gold. . . . Only one star was to be seen. It stood straight above Cape Chelyuskin, shining clearly and sadly in the pale sky. As we sailed on and got the cape more to the east of us, the star went with it; it was always there, straight above. I could not help sitting watching it. . . . Many a thought it brought to me, as the Fram toiled on through the melancholy night, past the northernmost point of the old world.[4]

Throughout much of his extensive voyage in the North Polar region, Nansen continually found himself confused and essentially lost. He stated:

It became more and more of a riddle to me that we did not make greater progress northwards. I kept on calculating and adding up our marches as we went along, but always with the same result; that is to say, provided only the ice were still, we must be far above the 86th parallel.⁵

The explorers later encountered muddy, warm water with much less salt which Nansen thought came from the Lena River in Siberia. However, they were still heading in a northwesterly direction. They sailed north in clear water at a good speed but were no nearer the pole than they were two weeks before. Upon hauling up water to wash the deck, Nansen noticed that it was sparkling with phosphorescence and crustacea were caught which looked like glowing embers. Nansen described one incident as follows:

... small crustacea and other small animals shining with such a strong phosphorescence that the contents of the net looked like glowing embers as I emptied them out in the cook's galley by lamplight.⁶

The Nansen expedition also encountered great quantities of dust which constantly blanketed the deck of their ship. Reed concluded that it came from volcanoes in the Earth's interior because it was found to be made of carbon and iron. In addition, the Fram constantly sailed against a current running south which baffled the crew because they expected a north-going current. Nansen commented on the incredible warmth experienced in the middle of December north of the 81st parallel and even found a warm-blooded animal at the 85th parallel. These findings seem to suggest that Nansen traveled into an area which should not exist in the North Polar region if the conventional theory about the Earth's shape is correct. As a result of his experiences, Nansen was convinced that the North Pole was located inside a very deep, ice-free basin or depression.

Other North Polar explorers observed similar phenomena which made no sense to them. Greely was confused when two of his men found a large, coniferous tree on a northern beach shore bordering the Arctic Ocean. Driftwood was seen all over. Reed concluded that this wood is washed into the Arctic Ocean from rivers flowing out of the Earth's interior.

Icebergs present another enigma if the ocean waters farthest north are clear of ice. One conclusion is that open water extends far enough into the interior of the Earth to escape the extensive cold experienced near the edge. Reed contended that North Polar icebergs were also the result of rivers flowing out of the interior. When these rivers reach the cold regions near the outer surface, they freeze and form icebergs. During the summer months, the bergs are thawed loose and washed into the ocean.

The modern polar expeditions of Admiral Byrd were military operations on a grand scale and a great deal of effort went into them. Byrd had essentially unlimited financing and modern technology to aid him. Airplanes could cover great distances in short periods of time, avoiding the difficult conditions experienced by earlier explorers. Dr. Raymond Bernard wrote a book entitled *The Hollow Earth,* published in 1969. In it, he quoted Ray Palmer, the late editor of *Flying Saucers,* as follows:

Admiral Byrd's two flights over both Poles prove that there is a 'strangeness' about the shape of the Earth in both polar areas. Byrd flew to the North Pole, but did not stop there and turn back, but went for 1,700 miles beyond it, and then retraced his course to his Arctic base (due to his gasoline supply running low). As progress was made beyond the Pole point, iceless land and lakes, mountains covered with trees, and even a monstrous animal, resembling the mammoth of antiquity, was seen moving through the underbrush; and all this was reported via radio by the plane occupants. For almost all of the 1,700 miles, the plane flew over land, mountains, trees, lakes and rivers.[7]

If the above account is true, then Admiral Byrd and those accompanying him explored a lot more than the public was ever told about.

It is possible that the Navy explored some of the interior world during their 1958 expedition under the "ice" to reach the "North Pole." For some mysterious reason, the mission was classified top secret and the public was not told anything about it until after the Nautilus submarine surfaced off the coast of Iceland on its return trip.

Additional information also points to the existence of these polar depressions. For example, in 1960 a picture was published in a Toronto newspaper, *The Globe and Mail,* which was taken by an aviator in the Arctic region showing a green valley containing rolling hills.[8] The Russians were evidently aware of the polar situation because all of their Sputnik satellites and manned space shots had orbits which missed the poles by a considerable distance. Finally, rumors have persisted that a number of early transpolar satellites mysteriously vanished over the North Polar region. Naturally, the gravitational variances would be horrendous, even if the satellites were thousands of miles above the Earth. At an orbital altitude of 100 miles, satellites might travel into the depression or crash on one side of it.

In summarizing this chapter, evidence for a hollow Earth with North and South Polar depressions or openings was presented. Satellite pictures were referenced of what appears to be a large North Polar depression or opening; however, pictures of a southern depression could not be located. In fact, good polar photographs of any kind are difficult to obtain. The North Polar photographs located by the author were always taken by satellites or other spacecraft in orbits near the equatorial plane which do not provide close-up detail. The accounts of explorers over the last 100 years provided additional evidence of the North Polar depression. NASA lunar seismic experiments provide evidence that the Moon may be hollow with a relatively thin shell. This in turn requires the Earth to be hollow as seismic disturbances and certain satellite pictures of the North Polar region may indicate. Even the geomagnetic evidence seems supportive. Finally, evidence presented in this chapter supports the original contention of a high lunar gravity and substantial atmosphere. However, polar openings or depressions may not exist since the evidence for them acquired thus far is still not conclusive. If they do not exist, then there are many enigmas which have not been explained by orthodox science. It is again appropriate to emphasize that the existence of the Moon's high gravity and substantial atmosphere does not depend upon polar openings or depressions. Other evidence presented stands on its own in support of the contention of the NASA-military cover-up regarding the true nature of the Moon.

CHAPTER 10

EVIDENCE OF EXTRATERRESTRIAL INTERFERENCE IN THE SPACE PROGRAM

The UFO subject has been abused by many writers who produce books about UFOs with seemingly little thought or hard analysis. Hopefully, information contained in the remainder of this book will help explain certain aspects of the UFO enigma in a more logical manner. This requires a better understanding of how the universe functions and man's position in it. Every phenomenon must have a rational basis, regardless of how bizarre it may seem to be. In every case, the law of cause and effect must hold. If NASA fabricated information, it will show up as a fabrication when it contradicts other valid occurrences. The sum total of all phenomena must boil down to the real picture of what happened.

Logic dictates that if UFOs occupied by intelligent beings are close by, they must have a purpose in being here. If these beings are around, actions taken by Earth people will probably be closely monitored by them. The human beings who live on the outer surface of the Earth have an extremely bad reputation for being killers. Historians can rationalize the wars based on this or that reason; however, from the standpoint of advanced civilizations which are not part of the outer Earth world, the Earth's history would be appalling. How could UFO occupants trust Earth nations to be benevolent when these nations have such war-filled histories and are developing weapons which are more and more powerful?

A great deal of information exists about Earth people who claim to have been contacted by "space" people. Many of these contactees have allegedly been told that the Earth is being carefully monitored. The primary reason given for the monitoring is so that they will be able to quickly curtail a nuclear war should it begin. If this is true, then missiles and spacecraft would be closely observed to ensure that they carried no nuclear warheads and did no damage to the Earth or Moon. Therefore, if UFO encounters or sightings were made by the astronauts, it should come as no surprise. This would be especially true during the Apollo missions if these people are stationed on the Moon. The present discussion will focus on the evidence of extraterrestrials inadvertently supplied by NASA and other lunar observers. Contactee encounters will be discussed in the next chapter.

Centuries ago, lunar observers noted transient phenomena which could not be easily attributed to natural causes. For example, small dome-like upswellings have appeared and disappeared on the Moon. In 1788, the astronomer Schroeter attributed these domes to the industrial activities of the Selenite (Moon) people.[1] Naturally, he was not taken seriously. However, over 200 of these white, circular domes have been observed and catalogued in current times.[2] These hemispheres have varied in diameter between an eighth and a quarter of a mile, with an average of one-quarter; and between 20 and 30 of them of similar size were seen clustered on the floor of Tycho Crater.[3] These particular domes cannot be attributed solely to rounded hills or volcanic upswellings. Their unpredictable appearance and disappearance indicates that they are intelligently created and movable structures.

Schroeter saw a shadow in the lunar Alps in 1788.[4] At first he saw a light, but after the region had been illuminated, a round shadow appeared where the light had been. Since the shadow was round, the object which produced it was detached and above the lunar surface. After 15 minutes, it supposedly disappeared. It is possible that Schroeter saw a large aerial object which provided the illumination to create its own shadow. Many bright, round spots have also been sighted inside craters such as Plato and the Sea of Crisis. They frequently appear dome-like and change their brightness from night to night.[5]

George Leonard supplied photographic evidence that the Moon is being worked with massive machines in his book, *Somebody Else Is on the Moon*. He suggested that damage done to the Moon's surface at one time is slowly being repaired. He claims to have photographic evidence which shows craters being worked on and perhaps mined. It is now known that the Moon is rich in valuable elements which could be used for many purposes.

The evidence of extraterrestrial monitoring of the space program apparently began with the Mercury missions and continued through Apollo 17. During Cooper's fourth orbit over Hawaii in 1963, he heard a weird voice transmission in an unintelligible language. The tapes were later analyzed and it was determined that the sounds were not those of any known foreign language on the Earth.[6] On the final orbit near Australia, he sighted a UFO from the space capsule.[7] It was also allegedly seen by over 200 people at the tracking station.[8]

Supposedly one or more UFOs were seen on each of the 12 Gemini missions. After the 1966 Gemini 9 mission was scrubbed due to interference with the radio hookup, NASA released a statement on television that UFOs or unknowns had been seen by astronauts on several occasions.[9]

White and McDivitt saw and photographed an egg-shaped, glowing, silvery object which moved above and below them. Five frames with a movie camera were taken as it flew by, and the film shows an egg-shaped object with a fan-like glow and a long tail of light. The Mission Control report stated that Command Pilot Jim McDivitt reported seeing another object in space which appeared to have big arms sticking out. The report also indicated that he took some motion pictures of the object, but had difficulty because of the Sun.[10]

It is significant that Gemini 7 encountered a UFO and many little particles traveling by the space capsule. The silvery UFO was not the rocket booster because it was seen in addition to the UFO.[11] NASA commonly attributed the particles observed in other missions to urine droplets or paint fragments peeling off the spacecraft.

John Glenn was the first astronaut to discover the "fireflies" which were observed quite frequently throughout the space program. After he had moved out of the night side of his first orbit, he glanced back through the window and thought that the spacecraft had tumbled because he saw "stars." Glenn soon realized that the spacecraft had not tumbled and that he was surrounded by yellowish-green, luminous particles which looked like "fireflies." They appeared to vary in size between a pinhead and three-eights of an inch and were eight to ten feet apart, evenly distributed through the space around the capsule. Each time the Sun came up, he observed these particles for about four minutes. Glenn stated the following about them:

> During the third sunrise I turned the spacecraft around and faced forward to see if I could determine where the particles were coming from. Facing forwards I could see only about ten percent as many particles as I had when my back was to the sun. Still, they seemed to be coming toward me from some distance so that they appeared not to be coming from the spacecraft. Just what these particles are is still subject to debate and awaits further clarification.[12]

In spite of Glenn's clear statement that these particles did not come from the spacecraft, orthodox authorities attributed them to flakes of material coming off the capsule.

Although more UFO sightings could be referenced involving the Gemini missions, the most valuable information came from the Apollo flights. In going around the Moon, Apollo 8 astronauts supposedly sighted a "disc-shaped" object, experienced a "blinding light", and heard "intolerably high-frequency noises" from their radio.[13] Later, they sighted the object again with more of the brilliant light and experienced "a wave of internal heat within the space capsule."[14] The spacecraft supposedly began to "pitch and yaw" before control was regained.[15] It is also significant that when the astronauts came around the eastern edge of the Moon, all the water in a radiator in the spacecraft's cooling system had evaporated and had to be replenished![16]

According to official accounts, Apollo 10 had a close call when Cernan and Stafford descended to within 50,000 feet of the lunar surface to inspect the Apollo 11 landing site. It seems that after the descent stage was jettisoned, the upper stage went into a wild spin, pitching up and down. Something caused the gyro-guidance system to go out of control, and to steady the ship, Stafford took over the hand controls. Supposedly, a control switch had been left in the wrong position by technicians and was not spotted by Stafford. However, official accounts do not mention that a UFO rose vertically from below and that it was photographed.[17]

The first Apollo 11 UFO encounter occurred one day into the flight. The astronauts sighted an unknown which appeared between them and the Moon which might have been the booster rocket. In the debriefing, Aldrin stated that they recalled having trouble with the high gain at about the same time as the sighting. Collins mentioned that they felt a bump and Armstrong indicated that Collins thought the MESA (Modularized Equipment Stowage Assembly) had come off. Aldrin then recalled seeing all sorts of little objects going by prior to seeing the brighter object which had an L-shape to it. Armstrong described it as an open suitcase and Aldrin later stated that they saw what appeared to be a cylinder. Armstrong compared it to two connected rings, but Aldrin disagreed and described it as a hollow cylinder. Collins re-entered the conversation and asserted that it looked like a hollow cylinder which was tumbling, but would change to an open-book shape.[18]

There are some significant items of information in the above conversation. First of all, Aldrin stated that they were having a problem with the high gain about the time of the sighting. Collins mentioned that they felt a bump, but then decided against this idea after Armstrong commented on the MESA package. As the dialogue progressed, the astronauts began talking about the shape of

Evidence of Extraterrestrial Interference 151

the object. It seems that these three, trained observers could not or would not agree on what they saw. Each man seemed to have a picture of what it was, yet after Collins explicitly described it as a cylinder, Aldrin stated that it was not a cylinder. Armstrong said that it looked like two connected rings.

When Apollo 11 was near the Moon, weird radio noises were heard which sounded like fire engines, sirens, buzz-saws, and train whistles. Mission Control supposedly asked if someone else was up there with them. These signals or noises came from outside the spacecraft and also allegedly persisted intermittently throughout the first few days of the flight.[19]

According to one undisclosed source, after Armstrong and Aldrin landed on the Moon, immense UFOs lined up on the far side of the crater and monitored the astronauts.[20] Another reference to this event appeared in the *National Enquirer,* September 11, 1979.[21] In the story, a former consultant to NASA claims that the event occurred but was covered up. According to the article, the encounter was common knowledge in NASA.

The reader is reminded that some of this information is limited and perhaps questionable concerning the subject matter of NASA UFO sightings. If the accounts were evaluated solely on their own, the case would not be totally proven by any means. However, in conjunction with all of the evidence which remains, and all that has been presented prior to this, the incidents are more believable.

Apollo 12 nearly experienced a complete electrical shutdown shortly after take-off. The ship appeared to be hit by bolts of lightning 36½ and 52 seconds after launch; but since there was no thunderstorm in the area, the event had to be explained from another standpoint.

Some people speculated that the rocket created an electrical conductor to ground from the ionized exhaust and that lightning discharged through the ship. However, one source claims that observatories in Europe reported two unknown objects in the craft's vicinity as it headed toward the Moon.[22]. One object appeared to be following Apollo 12 and the other was ahead of it. Both were blinking on and off rapidly. The following day the astronauts reported two UFOs or bogeys about 132,000 miles out, and during their conversation with Mission Control, one of the objects flew off at high speed.[23]. As Apollo 12 approached the Moon, strange sounds were allegedly received by Ground Control which did not come from the space capsule but from elsewhere. The noise was also supposedly heard by the astronauts and was described as static, whistles, and a constant beeping sound.[24]

The subject of fireflies has been mentioned several times. When the Apollo 16 astronauts were coasting toward the Moon, they entered a region of space swarming with these particles. However, NASA insists that they were flakes of paint which was applied to keep the craft from overheating in the Sun. But again, accompanying this occurrence, Mattingly reported that problems had arisen with the guidance and navigation systems. There was no attitude indication and the gimbal platform had locked. A manual realignment became necessary and the "snowflakes" traveling with the ship interfered with the star sightings. Typically, no one determined what happened; they merely suggested that another "electronic glitch" had occurred. Somehow, electrical transients were created in the electrical circuitry which created temporary malfunctions which later disappeared. Before the Apollo 16 descent,

Evidence of Extraterrestrial Interference 153

another bad circuit in the steering system controlling the main rocket engine occurred. This caused the engine bell to oscillate sideways.[25]

It appears that electronic glitches, UFO sightings, and luminous particles may be related phenomena. An examination of this relationship provides a great deal of information about what may have really happened to the astronauts. John Glenn saw luminous particles for about four minutes during every orbit when coming into the Sun. He clearly stated that they came toward the capsule from a distance. Glenn may have been observing the disintegration of certain particles which are emitted continuously by the Sun. Accumulated evidence indicates that they fill the space between planets and galaxies, and that their constituent particles probably consist of photons of light. When these particles disintegrate, photons are released. The nature of these particles may be intimately connected with the energies used to propel the UFOs which were allegedly seen by some of the astronauts. When UFOs came in close proximity to the NASA spacecraft, the concentration of these "firefly" particles may have increased tremendously. The particles seem to carry an electrical charge and penetrate readily through the material of a spacecraft. If they disintegrate when interacting with matter and other particles in space, the astronauts would see them inside and outside their ships. The effect of these particles on the spacecraft would be to cause overloading of electrical systems and electronic noise in the radio communication circuits. Since they may be emitted in high concentrations by UFOs, objects in the vicinity would be bathed in them sufficiently to cause the overloads and short-circuiting. For example, Apollo 10's guidance system could have been short-circuited by these particles from an approaching UFO.

The internal heat allegedly experienced by Apollo 8 astronauts could also be the result of these particles. If a UFO came close to the Apollo 8 capsule, a high concentration of these particles could have penetrated the ship releasing a significant amount of heat. The strange radio noises and erratic ship behavior could also be due to released electrons. The loss of water from the Apollo 8 radiator might also be attributed to the energy field since these same penetrating charged particles would readily cause evaporation. The close approach of a UFO could have caused the water in the radiator to rapidly boil off.

When the Apollo 11 astronauts sighted a UFO on their flight to the Moon, Aldrin indicated in the debriefing that they had a problem with the high gain. This follows the same pattern as the other incidents: the UFO energy field evidently caused radio interference. Collins claimed that they felt a bump. This could have happened if the ship were given a high enough concentration of these particles. No mention was made of the luminous particles in this instance, but they may have been observed by the astronauts. Since radio noises were heard intermittently throughout the first few days of the flight, UFOs may have been close by during much of their journey.

Apollo 12's electrical problems attributed to lightning may also have been caused by the UFOs allegedly spotted in their vicinity. Interestingly, the same radio noises and sounds were heard as on prior missions. If observatories noted that the UFOs were blinking on and off rapidly, it would not be surprising that the astronauts and Mission Control picked up a constant beeping

Evidence of Extraterrestrial Interference 155

sound. They were probably receiving the radio energy emissions of whatever caused the UFOs to blink.

During Apollo 16's Moon coast, the astronauts entered a region of space swarming with the "fireflies" or particles just discussed. NASA's explanation that the paint was flaking off was probably a continued effort to suppress the truth. If this paint was supposed to keep the craft from overheating and it came off early in the flight, it would represent shoddy engineering or workmanship. Since the same excuse was given for the fireflies on John Glenn's flight, no improvement in nearly ten years seems to have been made. The flaking paint provides a poor explanation for the phenomenon. NASA did not bother to explain why the paint always seemed to peel off when UFOs approached and why electronic glitches and other malfunctions also occurred simultaneously. At the same time Apollo 16 experienced the "paint" problem, they had guidance system problems. The energy field was so intense that it interfered with "star" sightings needed to realign the ship, so the Sun was used instead. Since electronic glitches seemed to occur most of the way to the Moon, the UFO must have been uncomfortably close.

Some of the problems that lunar probes experienced may have been caused in part by these same energy sources. Enough puzzling failures occurred that it seems possible that a number of Moon probes were tampered with. For example, one incident involved the Apollo 14 seismometer station which had been operating continuously from February 1971 to March 1975. The radio receiver went out in March 1975 and the transmitter stopped on January 18, 1976. The mystery occurred when both the radio receiver and transmitter began working about a month later on February 19, 1976. In addition, one of the other instruments which had never operated during the daytime began working perfectly night and day. Then, one month later the whole station quit working.[26]

The evidence that UFO occupants monitored and perhaps interfered with space program activities is substantial. It represents another item in a long list of the NASA-military cover-up. The reader is encouraged to pursue the referenced material for additional details and encounters not mentioned here.

In the next chapter, lunar information provided by UFO contactees will be examined and compared to what has already been presented. All of the available evidence must be explored if the truth is to be found.

CHAPTER 11

CONTACTEES AND MOON FINDINGS

George Adamski wrote a book with Desmond Leslie in 1953 entitled *Flying Saucers Have Landed*. It deals with historical UFO sightings and Adamski's initial experiences as a UFO contactee. Another book by Adamski in 1955, entitled *Inside the Space Ships*, covers his subsequent encounters with UFOs and their occupants. The following account was summarized from these two books.

Adamski was an amateur astronomer living in Palomar Gardens, California. In 1946, he observed a gigantic spaceship hovering above a mountain ridge near his home. This encouraged him to spend a considerable amount of time watching for them on a regular basis. In August 1947, he and four other observers counted 184 UFOs moving across the sky in about an hour's time. As a result of the great number of sightings which had occurred in this period and Adamski's reputation, some military people supposedly asked him if he would attempt to photograph them with his telescope. By 1951, he had successfully taken over 500 UFO photographs and became quite well-known for his work in the UFO field.

Reports indicated that a number of saucers had been seen landing in the desert area to the east. In an effort to make closer contact, he made a number of trips to the desert in 1951 and 1952. It was on November 20, 1952, 10.2 miles from Desert Center, toward Parker, Arizona, that he first made contact with a spacecraft and its pilot.

The craft landed and a man soon appeared, motioning Adamski to come over. In their conversation, the man indicated that one of their purposes for being here involved nuclear bombs and radioactive fallout.

Before sighting this smaller ship, Adamski and his three companions had seen a much larger, cylindrically-shaped craft. The space visitor indicated that the smaller ship came from the larger one. Adamski's three companions were then some distance away, waiting for him to signal them when he finished. He had decided that if contact were to be made, it might be best if only one person were involved, so his friends waited behind. At this initial meeting, Adamski did not enter the ship, but was able to carefully observe it from the outside. In less than an hour after the sighting, the space visitor said that he had to leave and the craft took off. Adamski then signaled his companions and they made sketches and castings of the visitor's strange footprints. They later reported the incident to the *Phoenix Gazette*.

On December 13, a small craft hovered above his house and he proceeded to take photographs of it. It approached within 100 feet and a film holder which Adamski had given to the man on the prior visit was dropped from one of the portholes. When the film was developed, one of the frames was replaced with a symbolic message.

Adamski was later given a ride in a small craft to a mother ship which was hovering about 40,000 feet above the surface of the Earth. It was 150 feet in diameter and 2,000 feet long according to one of the crew. On board, he saw a picture of a larger ship which was several miles long, the equivalent of a traveling city. The mother ship then moved out to a position 50,000 miles from the Earth, and Adamski described what he

Contactees and Moon Findings

saw through one of the portholes. He noted that the background of space was totally dark, yet it appeared that billions upon billions of fireflies were flickering everywhere.[1] They were moving in all directions and were of many colors, like a gigantic celestial fireworks display.

According to the space guide, the ship utilized the power of nature referred to as electromagnetic and had excess power at all times.[2] Some of the excess energy dissipated through its skin into space for short distances at times, but up to many miles on other occasions. The energy field acted as a shield, repelling particles and space debris with a constant radiating force. After a great deal of discussion, Adamski was returned to his house in the small "Scout" ship.

Two months later, Adamski was again contacted and taken to a different mother ship which was used as a laboratory for scientific analysis. These people explained to him that many of the smaller unmanned craft were used to collect data for their studies. Atmospheric samples were constantly collected and monitored for dangerous radioactive substances indicative of nuclear bomb testing. Adamski described one of the tests involving visual images of dust or space debris.[3] He observed the constant swirling activity of tiny particles of matter on a screen. Fine matter would appear to condense into a solid body at times, and then disappear, reverting to a nearly invisible state. The formations would occasionally become so rarefied that they seemed to be transmuted into pure gases. With the formation of particle bodies, certain quantities of energy would take visible, solid form, and then immediately dissipate in an explosion or disintegration that was visible on the screen. Another group of instruments recorded the intensity and composition. The cycle of formation and disintegration was ceaseless, involving energy and matter reacting with other space particles. Adamski noted that when the energy gathered into a sheet-like formation or a cloud-like body, it disturbed everything near it in space. He believed that he was observing the force that permeated all space, and from which planets and galaxies are formed, as well as the force supporting and sustaining life and activity in the universe. The guide indicated that the same power propelled their ships through space.

During this particular visit, the craft moved closer to the Moon and Adamski's guide informed him that the Moon had air as indicated by their instruments.[4] He told Adamski that air is not normally an obstruction to viewing another body as is claimed on Earth. He also mentioned the moving shadows of clouds which have been seen by Earth scientists on occasion. In addition, he said that the side of the Moon facing the Earth has clouds which are rarely heavy; however, the instruments indicated much heavier cloud activity, similar to Earth's, in the temperate zone just beyond the rim. He compared the near lunar side to the Earth's deserts, and mentioned that the temperature is not as hot as Earth scientists believe. He also stated that there is a strip of land around the center of the Moon which supports vegetation, trees, animals, and humans.

Adamski then described what he saw when looking through the ship's telescopic instrument.[5] He was amazed to discover how wrong Earth people were in their ideas about the Moon. Many of the craters turned out to be large valleys surrounded by mountains, and he stated that he could see definite indications that water must have existed on the near side at one time. His guide told him that there is still plenty of water on the opposite side and much hidden deep within the mountains on the near side. He also pointed out ancient waterlines on the flanks of mountains surrounding the craters and Adamski noticed deep ruts through the ground which must have been made by a heavy runoff of water. Adamski even saw vegetation and described the surface as fine and powdery, while other areas consisted of larger particles like coarse sand or fine gravel. As he watched, a small, four-legged, furry animal ran across the area he was observing.

On August 23, 1954, Adamski was taken on another trip to the Moon. He was shown large hangars in the bottoms of craters which were made to house large spaceships and was told that Moon visitors had to go through a depressurization process to acclimatize them to the atmosphere. It seems that this was necessary to keep them from experiencing the discomfort associated with high elevations and correspondingly lower air pressures.

When they reached the far side, the guide pointed out snow-covered mountains with heavy timber on the lower slopes.[6] They also observed mountain lakes and rivers which emptied into a large body of water. A number of communities were seen in the valleys and on mountain slopes as well as a fair-sized city. His guide indicated that hangars (domes) were constructed near the cities for convenience in landing with supplies which are brought in to exchange for certain Moon minerals.

The reader is reminded that Adamski's experiences occurred 15 years before men first landed on the Moon. Adamski's incredible observations will be appraised after those of another contactee are presented.

Howard Menger wrote *From Outer Space to You* in 1959. His experiences are just as shocking as Adamski's. In August 1956, he was taken to the Moon and observed dome-shaped buildings and a craft making a landing near one. On the next trip in September, the ship actually landed and he was allowed to take certain pictures. Each of the three pictures appearing in his book shows a bright, illuminated, lunar sky. Two of the pictures show the white, hemispherical domes which have been referred to previously.

Menger described the dome-shaped building as 150 feet in diameter and 50 feet high, made of translucent material. The dome rested on a pedestal made of a solid white substance. After leaving the spaceship which had entered the dome, Menger and the other Earth visitors were led to a train-like vehicle with plastic domes over each coach. The vehicle had no wheels and was suspended about a foot above a copper highway. They were soon gliding noiselessly, visiting building after building. Other landmarks and points of interest were the mountains, valleys, and underground installations. In a section of the Moon near the far side, Menger saw terrain which reminded him of the Valley of Fire in Nevada. At that point, the guide permitted the visitors to stick their heads outside for a moment. The heat was like a blast furnace and Menger stated, "I was certain no one could have lived outside very long . . ."[7]

Menger observed wind funnels forming on the ground, drawing dust into tiny whirlwinds. The sky had a yellowish color and the mountains jutted into a saffron-colored sky. He described the ground as yellowish-white, powdery sand with stones, boulders, and minute plant life here and there. There were other visitors from Earth, including Russians, Japanese, and Germans.

Contactees and Moon Findings

The experiences and observations of these UFO contactees will now be examined in light of the other evidence. On Adamski's initial trip to the mother ship, he observed space through a porthole when they were 50,000 miles above the Earth. He noted that the background of space was totally dark. This implies that stars are not visible above the atmosphere with the naked eye as suggested in Chapter 7.

Adamski also observed the firefly phenomenon and described it in the same manner as John Glenn did. How could Adamski have possibly discovered this in the early 1950's if his story about the trip were not true? It is important to note that the "paint flake" explanation for the firefly effect in space is probably a NASA fabrication. UFOs would not use paint as a shield from space debris or heat. The intensity of the firefly effect observed by Adamski may have been much greater than any seen by the astronauts. This would be expected if he were looking at space through the intense energy field of the spaceship.

The guide then explained the nature of the ship's propulsion source in terms that Adamski could understand. He stated that the ship's energy was radiated into space for short distances at times, but up to several miles on other occasions. This explains why the astronauts would observe the firefly effect if they were closely followed or approached by UFOs. The UFO energy field would also account for the radio interference and other electronic glitches which occurred on some of the NASA space vehicles.

On his second space ride, Adamski observed space dust being activated by the same energy which powered the ship. These energy particles seem to carry a negative charge and may be comprised of photons of light. Since space dust has a slight overall positive charge, these negatively charged particles would be attracted to the dust. Adamski may have been observing these negative charges building up on space dust particles until an excess accumulated. At this point, the dust particles seemed to explode and disappear because the negatively charged energy particles rapidly disintegrated. The cycle would then repeat itself. Similar phenomena were noted to a lesser extent in the Earth's atmosphere by the scientist Wilhelm Reich who called the particles orgone energy.

As the ship approached the Moon, the guide explained that the instruments were registering the Moon's atmosphere. He also explained that air is not normally an obstruction in viewing another body. This contention was presented in Chapter 7. The space guide then described the shadows of clouds which had been seen by Earth astronomers in lunar valleys and craters. He explained that they are rarely dense, but that in the temperate zone can become more so. In addition, he claimed that the temperature extremes are not as high as that predicted by Earth scientists, and that a strip of habitable land exists with vegetation, trees, animals, and people.

Adamski then looked for himself and described things which seemingly could not have been known before the space program unless his trip actually took place. He observed ancient watermarks on mountains and craters which were later discovered by the Apollo astronauts. In addition, he described deep ruts in the ground and was convinced that only a heavy runoff of water in times past could have caused them. This same conclusion was derived independently in Chapter 8 based upon the evidence supplied by NASA. Further-

Contactees and Moon Findings

more, Adamski stated that some of the surface appeared fine and powdery while other areas consisted of coarse sand and fine gravel. This sounds like Neil Armstrong's description of the lunar surface in the Sea of Tranquility. Adamski also saw sparse vegetation and a furry, four-legged animal running across his field of vision.

On the trip to the far side in August 1954, he was shown hangars which were used for housing large spaceships. He was also told that visitors went through a depressurization process. This would be expected in the higher elevations of the Moon where the air pressure would be substantially lower than in the ocean basins.

The guide pointed out snow on the peaks of the higher mountains on the far side of the Moon, as well as timber, lakes, rivers, and a large body of water. Adamski saw communities of varying sizes in valleys and on mountain slopes. A populated city came into view with people, buildings, and so on. There were hangars for landing and the guide explained that they exchanged supplies for certain minerals processed by the Moon inhabitants. NASA determined that the Moon is rich in many exotic metals. In addition, George Leonard provided evidence from NASA photographs that the Moon is being worked and mined. It seems that George Adamski somehow knew far more about the Moon than could ever have been observed through the most powerful telescopes on Earth. His books were written in the early and mid-1950's, and the Russians had not even put a satellite in orbit around the Earth at that time. His observations provide powerful evidence that he was telling the truth!

Howard Menger's alleged experiences are just as revealing. His photographs and experiences suggest that the Moon has a breathable, dense atmosphere. The white, dome-shaped building he photographed was probably the same type of building which Adamski described as a hangar. He described a vehicle with no wheels suspended above a copper highway. The copper highway could be indicative of a type of antigravity based on electricity. Believe it or not, U.S. patents already exist which may account for the principle of operation of this lunar vehicle! Some of this information will be presented in the chapter on gravity.

When Menger reached the terrain which reminded him of Nevada, he actually claimed to have breathed the lunar atmosphere. As expected, he held that the heat was extremely great, but the air pressure was evidently adequate to sustain life! He also observed wind, a yellow sky overhead, and a yellowish-orange sky near the horizon. The colors indicate that the atmosphere at that elevation might be deeper than the Earth's. As light passes through more atmosphere, it shifts to longer wavelengths. The yellow sky overhead indicates a longer wavelength than the blue in Earth skies. Similarly, the orange hue above the mountains can be attributed to light which is shifted in wavelength still more as a result of the increased thickness of the atmosphere when looking through it toward the horizon. Menger's description of the surface was similar to Adamski's: a yellowish-white, powdery sand with stones and miniscule plant life.

The observations of these contactees corroborate the author's findings using information obtained through space program activities. These men provided consistent, noncontradictory information. Most of their findings would have been difficult to surmise unless they actually visited the Moon. Since their books were

Contactees and Moon Findings 167

published over 10 years before Earth men first landed there officially, their stories are probably true. However, even without the supportive information supplied by UFO contactees, the evidence for a substantial lunar gravity and atmosphere stands on its own.

There are still many unanswered questions, but this is to be expected. The subject matter presented in this book has been suppressed to such an extent that the author could not readily walk down to the newsstand and read-all-about-it.

The next chapter uses the information presented thus far, plus some additional findings, to reconstruct the history of the Moon. Its history seems to involve many other planets in the solar system.

CHAPTER 12

MALDEK AND THE MOON

Evidence was presented in the chapter on Moon geology that many of the Moon's craters may not have been caused by volcanism or meteor impact, nor could they have been produced by nuclear weapons known to currently exist on Earth. It seems odd that the lunar maria are relatively devoid of extensive craters, especially since they cover such a large percentage of the lunar near side, while the far side has essentially no maria and is almost completely covered with craters. This uneven condition does not support a theory of random meteor bombardment. The cover photo of the Moon taken by Apollo 17 exemplifies this lack of randomness. The left half of the Moon is part of the near side, visible from Earth, while the right half is part of the extensively cratered far side. The seas or maria appear as dark areas on the left side.

As mentioned in a prior chapter, the maria are much lower in elevation than the rest of the Moon. In addition, there is evidence of what appear to be dried-up riverbeds flowing through valleys into these maria. All of this evidence suggests that the maria could have been oceans and seas before the lunar surface destruction. If they were, then they would be for the most part uninhabited, and in the event of a war, there would be no reason to bombard them. This then suggests that many of the Moon's craters, large and small, were the result of a war in which incredibly powerful weapons were aimed at specific, inhabited locations.

Craters have also been discovered on Mars, Mercury, and some of Jupiter's moons. If planets are fragile structures with relatively thin shells, a nuclear explosion or meteor capable of producing a crater 100 miles across would probably break open the shell of a planet the size of the Earth. Therefore, the Moon might have been reduced to rubble if all of the larger lunar craters were due to either of these causes. Many scientists believe that the asteroid belt is the debris of a planet which exploded. The debris moves in the same orbital plane as the planets and seems to be the correct distance from the Sun where a planet should be found. However, the explanation for what would cause the planet to explode still remains a mystery to orthodox science.

A planet could probably be reduced to rubble by focusing particle beams of sufficient strength to points deep within its shell. This would cause internal explosions that could rip the planet apart. Alternatively, a particle beam could be given a wide, diffuse pattern which would have minimum penetration. This might completely obliterate thousands of square miles of the surface and produce large, relatively shallow craters. If an interplanetary war occurred which involved Mercury and Mars, it seems likely that the Earth would have also been bombarded. Significantly, aerial photographs have shown remnants of large Moon-like craters along the eastern seaboard of the United States, in Northern Canada, and in other parts of the world.

The destruction of the asteroid belt planet, hereafter referred to as Maldek, might have generated some of the comets and meteors which are still in existence today. It is reasonable to conclude that the present asteroid belt comprises only a fraction of the debris from the explosion. Pieces of Maldek would have flown off in all

Maldek and the Moon 171

directions. Some would have escaped the pull of the Sun and the rest would have assumed a wide variety of eccentric orbits around the Sun in many different planes. Since the planets and the asteroid belt have the same orbital plane, the debris would intersect the orbital plane of the planets. This would pose a continual threat to the safety of every planet in the solar system.

If an interplanetary war took place within the memory of man, legends and records of the event might exist on Earth. Immanuel Velikovsky studied the historical texts, folklore, archaeological findings, sacred books, and literature of many cultures in search of evidence of an ancient catastrophe. From this exhaustive study, he concluded that cataclysmic events occurred approximately 3,500 and 2,600 years ago. The results were compiled in his book, *Worlds in Collision,* published in 1950.

Velikovsky surmised that Venus almost collided with the Earth 35 centuries ago. It looked like a comet and produced great tides that inundated large areas to great elevations. Its tail produced devastating effects as the Earth passed through it. During this encounter with Venus, the legends and accounts state that the Sun stood still for several days when the Earth stopped rotating. At the end of this period, the Earth began rotating again in the opposite direction. Prior to this, the Sun rose in the west and set in the east.

Velikovsky contended that in approximately 700 B.C., Mars created widespread damage by nearly colliding with the Earth. It also approached Venus and this allegedly caused both of them to move into their present orbits. Velikovsky's controversial ideas incited the wrath of the entire scientific community and he was viciously and unscrupulously ridiculed. Findings of the space program later tended to corroborate some of the statements he made. A book entitled *Velikovsky Reconsidered* goes into this in more detail.

Velikovsky also provided evidence from ancient accounts that the Moon has been an Earth satellite for a relatively short period of time. The reader is referred to an article written by Velikovsky entitled "Earth Without a Moon" which appeared in the book *Velikovsky Reconsidered*. In it, he cited references made by Aristotle, Democritus, Anaxagoras, Apollonius Rhodius, Plutarch, Assyrians, Indians, and even the Scriptures which mention the time when the Earth was without a Moon. It was in an early age but still within the memory of man. Therefore, it seems possible that the Moon may have appeared in orbit around the Earth after the destruction of Maldek. Perhaps it was Maldek's moon before the Maldek catastrophe. If so, it would have been heavily bombarded, also. Some of the Moon's extensive craters could have been the result of planetary debris colliding with it, but the absence of significant cratering in the maria and other evidence points to the aimed-weapon theory.

After Maldek was destroyed, Venus may have captured a great cloud of debris that made it look like a comet. However, it is difficult to explain why Mars would suddenly change its orbit and assume a new one that intersected the orbits of Earth and Venus, assuming purely natural causes. Another possible explanation for the cataclysmic events which Velikovsky discovered is the near approach of Maldek fragments to the Earth which may have been mistaken for Mars and Venus.

If the Moon recently moved into orbit around the Earth, and if the orbital changes of the Earth, Mars, and Venus really occurred as Velikovsky suggested, then

Maldek and the Moon

the moves almost had to have been intelligently executed. The Moon displays a number of bizarre orbital characteristics which cannot be easily explained away by coincidence. It orbits the Earth at exactly the right distance to make it appear to have the same diameter as the Sun. This is incredible in itself, but it is even more amazing that we never see the far side of the Moon. This is because it orbits the Earth every 28 days or so, and revolves on its own axis in exactly the same length of time. If these two periods were not the same, the far side of the Moon would have periodically come into view over centuries of observation! There is an important benefit which arises from the Moon's present orbital configuration. If the Moon rotated on its own axis such that both hemispheres faced the Earth periodically, the Earth's gravitational pull would generate devastating tidal forces on any bodies of lunar surface water. These orbital characteristics suggest that the Moon's placement in orbit around the Earth may have been an intelligently engineered project designed to minimize the Earth's tidal forces on it.

If Maldek were destroyed in a war, sophisticated weapons and incredible powers were probably available; therefore, gravity control must have been mastered by the civilizations involved. To accomplish the incredible planetary moving feats referred to above, giant spaceships with great levitating or gravity-inducing beams may have been used. The same devices could have been used to reverse the rotation of the Earth.

The orbital changes described in Velikovsky's book could have suddenly and drastically affected the Earth's geology and climate. Powerful evidence for these sudden changes is described in a later book by Velikovsky entitled *Earth in Upheaval*. For example, there are indications that areas that were tropical are now frozen wastelands. Frozen mammoths have been found with tropical plants in their mouths and stomachs. A six-month old, baby woolly mammoth was found deep-frozen in Siberia in 1977.[1] Scientists at the University of Rochester used a special technique to carbon date the animal and found that it had been buried for about 27,000 years. This date is amazingly close to astronomer Thomas Gold's suggestion presented in Chapter 8 that a flare-up of the Sun occurred less than 30,000 years ago and lasted for 10 to 100 seconds, scorching the Moon in the process. All of this might have been caused directly or indirectly by the destruction of Maldek.

It may be that a multiplicity of cataclysmic events occurred over a long period of time. Velikovsky's dates of 3,500 and 2,600 years ago seem to be too recent to account for the destruction of Maldek or for the placement of the Moon in orbit about the Earth. However, if planetary engineering is taking place on an as needed basis, perhaps Venus, Mars, and the Earth were moved at the approximate times Velikovsky suggested, causing the cataclysmic events he alluded to. Otherwise, the near approach of large Maldek fragments or meteors might account for the events. It is possible that a cataclysmic event took place 27,000 years ago which killed the woolly mammoth found in 1977. The climate could have been tropical in Siberia before that event, and similar events could have taken place 3,500 and 2,600 years ago. However, it would be expected that the woolly mammoths would have already been extinct by Velikovsky's dates due to prior catastrophes.

It seems as though the movie *Star Wars* had its basis in the Maldek story. The plot seems similar to the story in the Bible about the war in heaven involving Lucifer.

Interestingly enough, a particle beam or laser weapon was used in the movie to destroy a planet. The reader may or may not be aware that the United States and Russia are presently developing particle beam and laser weapons. It seems that the *Star Wars* plot could be much closer to historical reality than most viewers would have guessed. Although the Maldek story has not been constructed with any certainty, enough evidence exists to strongly suggest that this terrible war involving the Moon might have actually occurred.

CHAPTER 13

GRAVITY, SEARL DISKS, AND LEVITATING BEAMS

The probable nature of gravity and methods which UFOs could utilize for propulsion will be discussed in this chapter. It will be shown that an antigravity device may have been invented in 1949 which could have been used by NASA and the military to land men on the Moon. There have been many discoveries concerning the nature of gravity, but they have been ignored and suppressed by the world of orthodox science. Only a brief summary of a few of these discoveries will be presented.

A radiation theory of gravity was developed in the 1950's which seems to fit many of the observed facts concerning gravity; however, it may require modifications as new data becomes available. That theory will be presented in part in the following discussion. As explained in Chapter 8, the fundamental flaw in Newton's Law of Universal Gravitation seems to be the assumption that gravity effects have an unlimited ability to penetrate through any thickness of matter. Gravity displays many of the properties of light because gravity effects seem to be produced by a highly penetrating radiation in the electromagnetic spectrum. If follows that this radiation penetrates matter for limited distances before it is completely dispersed.

A laboratory experiment was conducted by Sir Henry Cavendish in 1798 to determine the force of attraction between solid metal spheres. This experiment determined the gravitational constant in Newton's gravitational equation which was then used to determine the mass of the Earth. Because of the limited penetrating ability of gravity radiation, a gross error was made in applying this constant to bodies of planetary size. One reason small bodies cannot simulate the gravity conditions in planets is because the attenuation of gravity radiation cannot occur to any comparable extent in small bodies. Another reason is a multiplication effect (not occurring in small bodies) caused by scattering of radiation into the gravity radiation band from higher frequencies. It is suggested that the matter of a planet below a certain distance from the surface will not contribute to the surface gravity in proportion to its mass because the gravity radiation generated by the matter will be partially dispersed or attenuated before reaching the surface. It follows that the mass of a planet cannot readily be determined without knowledge of the shell thickness, the percentage of caverns, and the average density of the shell material. The Moon would need an impossible mass to conform with the Law of Universal Gravitation because of its high surface gravity. Evidence has already been presented that both the Earth and Moon are hollow structures. Therefore, the mass of the Moon's shell is contributing more gravity radiation than that predicted by the Cavendish experiment.

The Sun emits radiation over a fairly wide spectrum, some of which must penetrate considerable distances into the planets. A minute percentage of the Sun's radiation must be in the gravity-inducing frequency band. The gravity radiation band is probably in the electromagnetic spectrum below the infrared radiation band. In penetrating large bodies like planets or the Moon, a significant amount of that penetrating radiation from the Sun between the infrared and the gravity band may be transformed or degraded into the lower frequency gravity radiation after passing through many miles of the bodies' shells. The transformation or multiplication effect would occur because of the scattering effects of atoms and molecules which gradually interact with the radiation and cause its average frequency to decrease until it degrades into the gravity band. This explains why bodies below a certain minimum size will not be able to transform any significant amount of radiation into gravity radiation. There simply is not enough matter in them to effect any transformation. This implies that the spheres used in the Cavendish experiment attracted each other with a smaller force proportionate to their masses than planets do. Therefore, the gravitational constant determined from the experiment was too small to predict the Moon's high surface gravity using Newton's Law of Universal Gravitation.

As mentioned previously, plumb bobs do not seem to be pulled out of line by mountains to the extent demanded by Newtonian Gravitation. This may be because mountains are not large enough to transform any significant amount of higher radiation generated inside them into gravity radiation, and the direct gravity radiation produced inside mountains may be partially dispersed by overlying masses. Therefore, the amount of gravity radiation coming from mountains would be even less than that expected from Newton's Law of Universal Gravitation.

The force of gravity accelerates every molecule of a small body almost equally and simultaneously; therefore, a free falling body experiences essentially no stress. If an intense gravitational field were used for UFO propulsion, it would readily account for their ability to make sudden right angle turns at extremely high velocities. No life form of appreciable size could withstand the accelerating G-forces associated with some UFOs unless an extraordinary means of propulsion were employed. To accomplish these feats, either the inertial mass of the system, including the occupants, has to be reduced to nearly zero, or every molecule needs to be given the same acceleration simultaneously.

Gravity-inducing radiation generators may be used by some UFOs for propulsion. It would explain how UFO contactees have been levitated into ships. However, a superior propulsion method would be to reduce the inertial properties of the ship and occupants to nearly zero if it is possible to do this. Inertia is defined as the ability to resist a change in velocity or motion. For instance, a large massive object has more inertia and is harder to move than a small one. Therefore, a ship with almost zero inertia could be given tremendous acceleration with a very small force, and the occupants would experience little, if any, stress. Several discoveries have been made which indicate that this may be possible.

A number of independent researchers, including Wilhelm Reich and Baron Karl von Reichenbach, demonstrated experimentally that certain types of negatively charged particles are repelled by a gravitational field. In the late 1800's, the scientist Sir William Crookes designed and performed rigorous tests on Daniel Homes, a man who could levitate himself and other objects. Crookes documented the experiments in articles published in the *Quarterly Journal of Science.* His book, *Researches in the Phenomena of Spiritualism,* contains these articles along with notes and professional correspondence. A book by Michael H. Brown that goes into recent experiments performed in levitation phenomena is entitled *PK, A Report on the Power of Psychokinesis, Mental Energy that Moves Matter.* The conclusion which may be derived from the experimental results produced by the above researchers is that gravity radiation consists of or is accompanied by particles which carry a negative charge. Gravity radiation and associated negatively charged particles will then attract solid matter because atoms and molecules seem to have a slight overall positive charge. It follows that a body containing a sufficiently high number of these negatively charged particles will levitate.

The above levitation phenomenon is apparently displayed by high altitude meteoritic dust. Dust particles below a critical size do not seem to fall in a vacuum. This is partially confirmed by meteoritic dust clouds 50 miles above the surface of the Earth. These dust particles remain suspended even though the density of the air is about 1/100,000 of the sea level value. In this case, the dust particles have accumulated a sufficent quantity of these negatively charged particles to levitate. When the negative charge disperses, the dust particles will fall.

The amazing Biefeld-Brown effect was discovered in 1923 by Thomas Townsend Brown and Professor Paul Biefeld of Denison University. These experimenters found that a parallel plate condenser charged with high voltage had a tendency to move in the direction from the negatively charged plate to the positive one, even in a vacuum. In other words, when this type of condenser is charged, an unbalanced or resultant force is exerted on it. Brown was granted patent numbers, 2,949,550; 3,022,430; and 3,187,206 in 1960, 1962, and 1965 for electrokinetic devices based on this effect. Over a period of years, Brown developed a number of experimental models with impressive propulsion capabilities. If the inertial properties of a spaceship could be reduced to nearly zero, the Biefeld-Brown effect might give it tremendous acceleration.

Evidence suggests that the inertial properties of a body can be greatly altered by impregnating it with energies of the right kind. Academic science recognizes only one kind of electron, but there is a great deal of evidence (such as that mentioned above) for the existence of electrons with much lower field intensities around them which are not easily detected or measured by conventional voltage and amperage meters. These electrons are probably the same particles associated with the propulsion systems of UFO spaceships mentioned in the chapter on contactees. They seem to accompany light and their constituent particles may be photons of light. Evidence supports the contention that these low field intensity electrons accompany photons emitted by the Sun and are responsible for all sorts of diverse phenomena. Since the Sun emits many different photons with different frequencies, the low field intensity electron spectrum would be equally broad. A body with a high enough concentration of these electrons will tend to levitate and be repelled by a gravitational field.

Although the exact particle nature of low field intensity, "subtle", electrons has never been understood, their effects have been recognized and called by different names such as kundalini, prana, mana, and the life force for thousands of years. The first truly scientific investigation into the nature of these subtle electrons was made in the mid-1800's by Baron Karl von Reichenbach at his castle near Vienna, Austria. He performed thousands of experiments over several decades and published a treatise in 1850 entitled *Researches on*

Magnetism, Electricity, Heat, Light, Crystallization, and Chemical Attraction in Their Relations to the Vital Force. Reichenbach termed the subtle electrons "Od", derived from the Norse deity Odin, indicating a power permeating all of nature. His exhaustive and precise research enabled him to discover the electrical nature of Od and he was able to show that it was closely associated with light. In addition, he determined that Od was associated with essentially all living organisms and nonliving physical phenomena such as chemical reactions, magnets, electricity, crystals, water and so on. He established that it behaved differently than normal electricity with different rules of conductivity through materials.

In 1908, Walter Kilner studied the characteristics of what he termed the "human atmosphere" using special filtering screens to make the aura visible. His findings corroborated Reichenbach's research concerning subtle electrons or the energy field emanating from the human body. Kilner's book, *The Human Atmosphere,* was published in 1911 and the revised edition entitled *The Human Aura* was republished in 1965.

In 1925, a Russian cytologist, Alexander Gurwitsch, detected the same electrons emanating from cells and called the radiation mitogenetic rays. He found that the radiation could be reflected, absorbed, and that he was able to increase the multiplication of yeast cells with it.

In 1936, Otto Rahn, a professor of bacteriology at Cornell University, published *Invisible Radiations of Organisms.* His findings concerned the same subtle electrons of Reichenbach, Kilner, and Gurwitsch.

In 1939, H.S. Burr and F.S. Northrop published their research findings concerning an electrodynamic or L-field emanating from organisms. The findings appeared in the *Proceedings of the National Academy of*

Science in an article entitled "Evidence for the Existence of an Electro-Dynamic Field in Living Organisms."

Dr. Wilhelm Reich was a prolific researcher into the nature of subtle electrons in the 1930's, '40's, and '50's. He termed these subtle electrons "orgone" because of their intimate relationship to living organisms. He was able to concentrate orgone in specially designed orgone accumulators which he used for healing purposes. This application led him into serious trouble with the medical establishment, even though he was a Doctor of Medicine. His exhaustive research established the relationship of orgone or subtle electrons with living organisms, healing energies, disease, atmospheric and weather phenomena, sunlight, radioactive substances, electricity and conductivity, visible color effects, heat, and many other things.

The above brief discussion has laid some of the groundwork for a better understanding of the Searl effect which could probably be utilized for space travel. The Searl effect was supposedly discovered by an English electronics technician by the name of John Searl in 1949.[1] He noticed that a small voltage was induced in rotating metal objects. The outer edge had a negative charge while the central region had a positive charge. He reasoned that free electrons were thrown outward from the center by centrifugal force. It followed than an electrical generator might be constructed based on this principle.

Searl's first generator consisted of a segmented, rotating disk with stationary electrodes used to collect electrons that were thrown out. The armature was three feet in diameter and was turned by a small engine. The results were completely unexpected. At low armature speed, the disk produced powerful electrostatic effects

Gravity, Searl Disks, and Levitating Beams

on nearby objects accompanied by crackling sounds and the smell of ozone. Beyond a critical rotational speed, the armature continued to accelerate without the aid of the engine. The generator finally lifted off the ground while still accelerating and rose about 50 feet, breaking the connection between it and the engine. It hovered at this height for a moment as its rotational velocity continued to increase during which time a pink halo appeared around it. This halo seems to be indicative of ionization of the surrounding atmosphere. Another effect was that local radio receivers went on of their own accord from the effect of the energy field surrounding it. Finally, it reached another critical rotational velocity, rapidly gained altitude, and supposedly disappeared from sight.

Since 1952, Searl and others have supposedly constructed generators of varying sizes from 3 to 30 feet in diameter. His amazing discovery would indicate that some UFOs work in a similar manner since many have been sighted with rotating segments. One of the most significant aspects of this invention is that at extremely high potentials, the component parts seem to experience a considerable loss of inertia. The authenticity of this information on Searl disks cannot be verified; however, assuming it is true, a logical explanation will be presented of how Searl disks probably work.

When the armature of the Searl disk rotated, normal electrons (those detected by standard instruments) were probably thrown outward toward the rim, generating the small voltage detected. However, they alone do not seem to be able to account for the subsequent phenomena. The subtle electrons which permeated the disk material would have also been thrown outward by rotation. Since subtle electrons seem to be very unstable, they will disintegrate with little provocation, releasing light and heat in the process. The rotating armature could have caused many of these subtle electrons to disintegrate. This in turn may have caused greater quantities of normal electrons to be released. In turn, powerful electrostatic effects would have been created around the armature.

As subtle electrons were thrown outward toward the rim, a temporary void was probably left in the central region of the disk. Additional subtle electrons from the surrounding areas would then move in to fill the void. This could have generated a vortex motion which increased the rotational velocity of the disk. In turn, as the velocity of spin increased, greater numbers of particles and regions of space probably became involved. The disk finally acquired such a strong negative charge that the Earth's gravity may have repelled it upward. The highest concentration of subtle electrons is probably close to the Earth's surface; hence, when the disk rose above this concentration, it may have lost some of its charge temporarily, causing it to hover. Finally, it could have again acquired a heavy negative charge by the same process, causing it to rise at a tremendous rate due to its reduced inertia. If the inertia of a spaceship could be nullified by the Searl effect, then the Biefeld-Brown effect might be used to give it incredible acceleration and velocity.

Searl disks and some UFOs seem to have many characteristics in common. In both cases, claims have been made that a halo or ionized region has been seen around them and that they have been observed to burn vegetation when hovering close to the ground. They both allegedly create electronic interference and noise effects. The ultrahigh concentration of subtle electrons around

Gravity, Searl Disks, and Levitating Beams 187

UFOs produced by the Searl effect would probably be injurious if a person got too close to one. UFO contactees such as George Adamski were warned not to touch them. This is probably because the subtle electrons surrounding UFOs are always in the process of disintegrating. A high concentration could enter the body and disintegrate, releasing great quantities of normal electricity which would cause internal and external damage. This is what may happen to vegetation which comes in contact or close proximity with them!

Another suppressed space program discovery which provides clues about the nature of gravity is termed "missile weight loss in space." A wide variety of subtle electron concentrations may exist in space, such as in the Van Allen Radiation Belts, which would cause satellites to lose weight and inertia. An account of satellites which lost weight and inertia was given by Frank Edwards in his book *Strange World*.[2] Edwards claimed that this phenomenon was officially confirmed by an Air Force statement made in November 1960. An undamaged Discoverer satellite, that had originally weighed 300 pounds before its polar orbit, was reduced to only 125 pounds after recovery a few days after launch. Another account given by Edwards referred to fragments of Soviet Sputnik 4 which exploded in orbit when its retro-rocket malfunctioned. The metal fragments were found to be extremely light. Shortly thereafter, additional fragments were found and tests demonstrated that the metal had lost more than half its normal weight. Even the jar in which one piece was placed began losing weight! This suggests that many of the subtle electrons captured by the satellite fragment escaped and were in turn captured by the container, causing it to lose weight, also. This discovery adds support to the theory of gravity radiation presented in this chapter and to the related theory of "subtle" electrons.

A number of startling discoveries presented in this chapter may have been used for landing men on the Moon. After all, these findings are more than 20 years old. Since the military has always been on top of new discoveries of this nature, it would be naive to assume they were not experimenting with antigravity devices long before 1969. Evidence for this will be presented in the next chapter.

CHAPTER 14

THE PRESENT AND FUTURE SPACE PROGRAM

Apollo lunar lift-off pictures provide evidence that rocket propulsion was not used to escape from the Moon's surface. The reader is referred to a reproduction of a color TV transmission of the Apollo 17 lunar lift-off in Photo 18. There are no signs of a rocket exhaust stream in this picture! A great deal of evidence has been provided that the Moon has a high gravity and a substantial, Earth-like atmosphere. The black sky in this picture indicates that the sky was probably filtered out. At first, it might seem that a darkening process used to blacken the sky also eliminated the exhaust stream. But if the sky were darkened, an atmosphere exists; hence, a high gravity exists and rockets would not work in the first place. A close examination of the photo discloses that the lunar ascent stage appears to be illuminated. This glow or illumination follows the contour of the Module and is clearly not the exhaust. However, it may be the reflection of the Sun off the side of the LM. If exhaust were apparent at all, it would be seen below the illuminated ascent stage.

Compare the above Apollo 17 photo with the Apollo 16 lunar lift-off shown in Photo 19. The red, blue, green, and yellow splotches seem to indicate that a lot of acitivity was taking place at the base of the ascent stage. One possible explanation is that explosive bolts were detonated which had connected the descent stage to the ascent stage. This may have caused metal fragments and other debris to be blown out from the base of the craft. Another possibility is that a small initial supply of rocket fuel was burned to convince the public that rockets were the sole propulsion means. The author obtained the 8mm film of the complete lift-off sequence and noted that the initial blast appeared as a red-colored plume coming out of the rocket nozzle, but stopped soon after the ascent stage separated from the descent stage.

Photo 20 is a picture of the Apollo 16 lunar lift-off an instant after Photo 19 was taken. The absence of a visible exhaust stream coming out of the rocket nozzle is very apparent. Those who attempt to discredit these NASA photos will probably assert that the exhaust would not be visible in a vacuum. However, chemical rockets eject a large quantity of combustion products at thousands of degrees Fahrenheit. These products and gases emit an intense amount of light in an exhaust stream which extends many feet down from the rocket nozzle. As the distance away from the nozzle increases, the gases and combustion products begin to disperse. The amount of light is often sufficient to illuminate the surrounding area, and the greatest light intensity is in the exhaust stream itself. A vacuum would have little effect in eliminating the emitted light from the main exhaust stream since the exhaust gases and other combustion products supply their own radiant energy or light. In any event, the initial blast or burn at separation (as explained in the previous paragraph) provides proof that these combustion products would have been visible for the duration of the ascent since they were clearly evident at the beginning. Perhaps the reader recalls artistic renditions of the lunar lift-off shown in newspapers and

The Present and Future Space Program 191

other literature prior to the Apollo 11 mission. An intensely visible exhaust stream was always shown coming out of the rocket nozzle. In summary, the photographs and films provide evidence that rockets were not used to escape from the Moon.

The aerodynamic problem of the Lunar Module has not been discussed, but it was clearly an aerodynamic disaster. NASA stressed that this was not a concern in the vacuum condition on the Moon. Since evidence has been supplied that the Moon has a dense atmosphere, the Lunar Modules would have been aerodynamically unstable if they attained high velocities. This means that the velocity must have been kept below a critical speed. This cannot be done efficiently with rockets because the fuel requirements become astronomical when the speeds are kept down. It is also significant that the astronauts were standing up during descent and ascent. Supposedly, they were held in place by restraining straps attached to the ceiling. Even so, no appreciable deceleration or acceleration could have been tolerated by the astronauts with this kind of arrangement. This means that the acceleration and velocity were kept low, creating the worst possible conditions for efficient use of rocket fuel. Just what method did NASA really use to take astronauts to and from the lunar surface?

Plans for the massive Saturn launch vehicle were undoubtedly on the drawing boards prior to the discovery of the high lunar gravity. Wernher von Braun had envisioned such a rocket many years before NASA was created. As large as this rocket was, one at least seven times larger would have been required to send a Lunar Module, which used chemical rockets, to the Moon. After antigravity devices were developed, the Saturn rocket would not have been needed. However, the interests of big business and the military were deeply involved in the space program. If the project continued, the military could continue to keep the new discoveries secret, and billions of dollars were to be made by big business and spent on secret military research projects. Once the astronauts were in space, advanced antigravity devices could be used in secrecy and the public could be kept in the dark.

The military provides the incentives or demands for expensive defense appropriations and big business stands to profit from the expenditures. The U.S. Government may be little more than a puppet for these groups. This should be apparent to the reader when considering that energy devices like the ones referenced in the last chapter could make the energy crisis disappear practically overnight. In actuality, the energy crisis is maintained for the profit of world energy cartels.

When NASA discovered that rockets alone were not feasible or adequate for landing men on the Moon, funds were undoubtedly channeled into gravity research and related projects. The military had probably already been working with the Searl effect and the Biefeld-Brown effect in the early 1950's and they may have perfected some of these devices by the mid-1950's. In addition, a device for generating gravity-inducing radiation may have been developed before 1960. It is a definite possibility that the United States military landed men on the Moon long before Apollo 11!

The new levitating devices used in the Apollo ships were not necessarily relied upon to the extent that the space vehicles could be completely maneuvered and controlled by these means. It is possible that they were only used as the main propulsion source for braking, soft-landing, and ascent. Attitude control might still have been accomplished with small thrusters. It would be much easier to maintain secrecy with the minimal use of the new antigravity devices.

Electrical energy is still required to power the levitating mechanisms. The Searl effect would only require a small amount of electrical energy to get the levitating process started and then energy from the surrounding space would sustain it. However, gravity-inducing radiation generators would require a continuous source of electricity. In any event, various combinations of Searl generators, Biefeld-Brown devices, gravity-inducing radiation generators, or other secret devices could have been used on the Apollo missions.

It may be that NASA and the Air Force have remained silent about UFOs because of their own development and use of antigravity devices. It is well-known that the Air Force put a lot of time and effort into studying UFOs as far back as the early 1950's. It would be naive to assume that they did not learn a great deal about them in the process. Their policy of total silence concerning UFOs indicates that they may have something to hide from the public. The author highly recommends a book entitled *The Roswell Incident* by Charles Berlitz and William L. Moore for a fully documented account of a government cover-up involving a crashed UFO in Roswell, New Mexico in 1947. It is entirely possible that many of the UFOs sighted in the 1950's and '60's were owned by the United States Air Force!

Russia's role in the whole space race affair has been very mysterious, to say the least. When they were on the verge of beating the United States to the Moon, they pulled out of the race. Or did they? Russia soft-landed Moon probes before the United States. This means that either antigravity was employed or the Moon's atmosphere was used to slow down the probes. If the lunar atmosphere were used, ablative heat shields similar to the ones used on space capsules for entering the Earth's atmosphere were employed. On September 20, 1970, Russia soft-landed Luna 16, a remote-controlled lunar probe which returned to Earth with a soil sample taken from the Sea of Fertility. This accomplishment was deemed insignificant in comparison to the Apollo Moon landings. However, the device probably would have required an antigravity propulsion system in the high lunar gravity. Since the United States' Surveyor probes did not carry enough fuel to soft-land in the high lunar gravity, it seems probable that antigravity was used for Moon probes, beginning in the mid-1960's, by both the United States and Russia.

Russia may have used antigravity devices to land men on the Moon and bring them back in total secrecy. After they learned in 1959 that the Moon had a high surface gravity, they knew it was impossible to succeed with rockets alone. They probably then focused their attention on antigravity, letting the United States put on the show for the world. By continuing to maintain a limited, rocket space exploration program, the secrecy of their antigravity exploration attempts could be maintained.

An article entitled "Russians Press to Dominate Space Again" appeared in the September 30, 1979 edition of *The Oregonian* newspaper.[1] The writer of this article referenced a message delivered to Congress by the former Apollo 10 astronaut, Lt. Gen. Thomas P. Stafford, who was the Air Force's deputy chief of staff for research and development at the time. Stafford held that the Russians were rebounding from the Moon race setback and were aggressively pushing their efforts as a sign of national power. They were supposedly working toward space stations, colonies, orbiting factories, and sending humans to other planets.

On October 11, 1980, two cosmonauts returned to Earth after setting a record of 185 days, or over six months, in the Salyut 6 space station. According to the above referenced article, Stafford believed the Russians were developing a manned, military space capability. Dr. Charles S. Sheldon II was also mentioned in this article. He indicated that the Russians might have begun using a totally new ferry craft. In addition, he anticipated that they would be building a large space station and a manned, interplanetary expedition force to explore other planets. However, Sheldon did not explain the propulsion devices the Russians would be using to explore them. It is economically prohibitive for any country to land astronauts on a planet with a high gravity using conventional rockets. Therefore, if Sheldon's contention is true, Russia must also have antigravity devices.

The reader must be reminded that the latest technology available to the military is always years ahead of publicized capabilities. Russia is probably no further ahead in developing levitating devices or exploring other planets than the United States. In all probability, the United States manned exploration of space never stopped after the Apollo missions. It is an interesting possibility to consider that Russia might be sending cosmonauts on extended trips to the Moon or other planets for up to six months using the Salyut 6 space station as a cover. If so, what might the United States be doing, and could the Space Shuttle Project be another decoy for military space operations?

As a result of the NASA cover-up, all findings concerning other planets released to the public by NASA and orthodox scientists are to be viewed with suspicion. The Moon's high gravity necessitates drastic revisions to cherished concepts of planetary gravity and atmosphere, as well as other cosmological concepts. The above considerations and a great deal of additional evidence suggests that a number of planets and moons in the solar system may be inhabited by intelligent people with technologies superior to ours. If so, the operations of the United States and Russia in space may be restricted by these people and limited to certain areas. We can hope that these people will keep the superpowers in check, since the use of space for military purposes could eventually lead to our destruction.

The evidence presented in this book indicates that a monstrous suppression of vital energy-related discoveries has occurred. The NASA cover-up is just a small part of it. The reasons for the suppression have been touched upon briefly, but the ramifications of the new findings are astounding. Consider what would happen if the United States Government issued an official announcement of these findings. A common knowledge of the nature of gravity and inexpensive ways to control it would revolutionize transportation. A new world with practically unlimited energy could emerge out of the chaos.

APPENDIX A

DERIVATION OF THE NEUTRAL POINT DISTANCES FROM THE MOON USING NEWTON'S LAW OF UNIVERSAL GRAVITATION

Since the forces of attraction of the Earth and Moon at the neutral point are equal, a generalized equation will be developed. The following parameters are used:

M = mass of an object at the neutral point
M_e = mass of the Earth
M_m = mass of the Moon = $M_e/81.56$
X = distance from the Earth's center to the neutral point
Y = distance from the Moon's center to the neutral point
R_e = radius of the Earth = 3,960 miles
R_m = radius of the Moon = 1,080 miles
G = gravitational constant
F_e = force of attraction of the Earth
F_m = force of attraction of the Moon
T = center to center distance between the Earth and the Moon.

The gravitational forces of the Earth and Moon on an object at the neutral point, using Newton's Law of Universal Gravitation, are as follows:

$$F_e = \frac{GM_eM}{X^2}$$

$$F_m = \frac{GM_mM}{Y^2}$$

Since $F_e = F_m$ and $M_m = M_e/81.56$,

$$\frac{GM_eM}{X^2} = \frac{GM_eM}{81.56Y^2}$$

Since $X + Y = T$, $X = T - Y$. Substituting this expression for X yields

$$\frac{GM_eM}{(T - Y)^2} = \frac{GM_eM}{81.56Y^2}$$

Canceling terms results in

$$\frac{1}{(T - Y)^2} = \frac{1}{81.56Y^2}$$

$$81.56Y^2 = (T - Y)^2$$

$$9.031Y = T - Y$$

$$Y = \frac{T}{10.031}$$

This is the generalized equation of the distance from the neutral point to the center of the Moon, Y, as a function of the total Earth to Moon center distance, T.

The distance between the Earth and Moon varies between 221,463 and 252,710 miles with an average of 238,885, assuming that these are the center to center distances. The corresponding neutral point distances from the Moon's center for T = 221,463; 238,885; and 252,710 miles are 22,078; 23,815; and 25,193 miles, calculated from the generalized equation for Y.

APPENDIX B

DERIVATION OF THE MOON'S SURFACE GRAVITY BASED ON THE 43,495-MILE NEUTRAL POINT DISTANCE

The following parameters are used:
R_e = radius of the Earth = 3,960 miles
R_m = radius of the Moon = 1,080 miles
X = distance from the Earth's center to the neutral point = 200,000 miles
Y = distance from the Moon's center to the neutral point = 43,495 miles
G_e = Earth's surface gravity
G_m = Moon's surface gravity

Since the forces of attraction of the Earth and Moon are equal at the neutral point, the inverse-square law yields:

$$G_e \left(\frac{R_e^2}{X^2} \right) = G_m \left(\frac{R_m^2}{Y^2} \right)$$

$$\frac{G_m}{G_e} = \frac{R_e^2 Y^2}{R_m^2 X^2}$$

$$= \frac{(3,960)^2 (43,495)^2}{(1,080)^2 (200,000)^2}$$

$$= .64$$

Therefore, $G_m = .64 G_e$

APPENDIX C

DERIVATION OF THE FINAL VELOCITY REACHED AT THE MOON BY A SPACECRAFT TRAVELING 2,200 MILES PER HOUR AT THE 43,495-MILE NEUTRAL POINT

The final velocity can be determined by deriving an equation based on the conservation of kinetic and potential energy. Basically, the kinetic energy of the spacecraft at the Moon equals the kinetic energy at the neutral point, plus the gain in energy from the Moon's pull, less the loss in energy from the Earth's pull. The equation is

$$KE_m = KE_n + P_m - P_e$$

where

KE_m = kinetic energy at the Moon
KE_n = kinetic energy at the neutral point
P_m = gain in energy from the Moon's pull
P_e = loss in energy from the Earth's pull

Each energy term is defined as follows:

$KE_m = \frac{1}{2} M V_m^2$
$KE_n = \frac{1}{2} M V_n^2$

$$P_m = \int_{1,080}^{43,495} MG_m \left(\frac{R_m}{x}\right)^2 dx = MG_m R_m^2 (.0009029)$$

$$P_e = \int_{200,000}^{242,415} MG_e \left(\frac{R_e}{x}\right)^2 dx = MG_e R_e^2(.000000870)$$

M = mass of the ship
G_m = surface gravity of the Moon = $.64(79,036)$
      = $50,583$ miles per hour^2
G_e = surface gravity of the Earth = $79,036$ miles per hour^2
R_m = radius of the Moon = $1,080$ miles
R_e = radius of the Earth = $3,960$ miles
x = distance from the Earth or Moon in miles
V_m = velocity at the Moon in miles per hour
V_n = velocity at the neutral point = $2,200$ miles per hour

Substituting the energy terms in the general equation results in
$$\tfrac{1}{2}MV_m^2 = \tfrac{1}{2}MV_n^2 + MG_m R_m^2(.0009029) - MG_e R_e^2(.000000870)$$

Canceling terms yields
$$V_m^2 = V_n^2 + 2G_m R_m^2(.0009029) - 2G_e R_e^2(.000000870)$$

Substituting known parameters results in
$$V_m^2 = (2,200)^2 + 2(50,583)(1,080)^2(.0009029) - 2(79,036)(3,960)^2(.000000870)$$

and finally,
$$V_m = 10,451 \text{ miles per hour}$$

APPENDIX D

DERIVATION OF THE COMMUNICATION BLACKOUT TIME BASED ON A MOON SURFACE GRAVITY EQUAL TO 64 PERCENT OF EARTH'S

The centrifugal force equals the force of gravity on an orbiting body. Therefore,

$$\frac{V^2}{R} = G_m \frac{(1,080)^2}{(1,150)^2}$$

at an altitude of 70 miles. Since R is 1,150 miles, the orbital velocity, V, can be determined if the surface gravity is known. G_m = 50,583 miles per hour² if 64 percent surface gravity is assumed. Substituting these values yields

$$V^2 = (1,150)(50,583)\frac{(1,080)^2}{(1,150)^2}$$

or

$$V = 7,163 \text{ miles per hour}$$

At an altitude of 70 miles, the ship will be out of communication with the Earth during 39 percent of its orbit. This is because it is high enough above the surface of the Moon to be visible to the Earth even though it has traveled beyond the surface horizon as seen from the Earth. After it passes an angle of arc cos $\frac{1,080}{1,150}$ or .939 past the halfway point of its orbit as seen from Earth, it will be behind the Moon. Arc cos .939 equals about 20°. Therefore, during (180° − 2(20°)) or 140° of its orbit, it will be in a communication blackout.

The fraction of its orbit during the blackout is $\frac{140°}{360°}$ = .389. A complete orbit at 7,163 miles per hour would require about 60 minutes. Therefore, the blackout time would last about 24 minutes.

APPENDIX E

DERIVATION OF THE LUNAR MODULE FUEL REQUIREMENTS IN HIGH MOON GRAVITY CONDITIONS

The chemical energy required to brake a spacecraft or send it into orbit depends on the gravity, exhaust velocity, time of burn, orbital velocity, and the payload. Fundamentally, the spacecraft must be lifted to a certain height and be given a certain orbital speed. The calculations which follow are for the ascent stage. The descent fuel requirements will be based on the payload ratio determined for ascent. Both payload ratios are nearly the same because approximately the same amount of energy must be expended to orbit or soft-land a pound of payload.

A vertical rocket with a given amount of fuel will burn out at a certain altitude and then continue to rise until its kinetic energy is reduced to zero by the constant pull of gravity. In order to send a spacecraft from the lunar surface into orbit around the Moon, it must be lifted to a certain height and then be given a horizontal velocity. If the Moon had a gravity equal to Earth's, the orbital velocity at 9.5 miles above the surface would have to be

$$\frac{V^2}{R} = 79{,}036 \left(\frac{1{,}080}{1{,}089.5}\right)^2$$

$$V^2 = 79{,}036 \, \frac{(1{,}080)^2}{(1{,}089.5)}$$

$$V = 9{,}197 \text{ miles per hour}$$

With a gravity 64 percent of Earth's, the orbital velocity at 9.5 miles would be

$$\frac{V^2}{R} = 50{,}583 \left(\frac{1{,}080}{1{,}089.5}\right)^2$$

$$V^2 = 50{,}583 \frac{(1{,}080)^2}{(1{,}089.5)}$$

$$V = 7{,}359 \text{ miles per hour}$$

The equation for the maximum height achieved by a vertical rocket after burnout is given as follows:

$$H_{max} = \frac{U^2(\ln R)^2}{2G} - UT \left(\frac{R}{R-1} \ln R - 1\right)$$

where,

R = payload ratio of the rocket
U = exhaust velocity = 6,312 miles per hour
T = rocket burn time = .121 hours
G = Moon surface gravity

The fuel used on the Lunar Module was a 50/50 mixture of hydrazine and unsymmetrical dimethylhydrazine with a nitrogen tetroxide oxidizer. The average exhaust velocity was 9,258 feet per second, or 6,312 miles per hour. The Moon's surface gravity at 64 percent of Earth's is 50,583 miles per hour^2. The rocket burn time for ascent was assumed to be the same as claimed by NASA, or 7 minutes 15 seconds. This is .121 hours.

The maximum height attained is calculated as follows: The kinetic energy required is equal to the difference in potential energy between the maximum height achieved and the height at burnout.

$$MGH_{max} = MGH_{burnout} + \frac{1}{2}MU^2_{burnout}$$

where,

Appendix E

$U_{burnout}$ = velocity of rocket at burnout
M = mass of rocket
$H_{burnout}$ = height at burnout

Canceling terms yields

$$H_{max} = H_{burnout} + \frac{U^2_{burnout}}{2G}$$

H_{max} will now be derived for the case where Moon gravity is 64 percent of Earth's. The following values are used:

$H_{burnout}$ = 9.5 miles (the value claimed by NASA)
$U_{burnout}$ = 7,359 miles per hour
G = 50,583 miles per hour^2

Substituting yields

$$H_{max} = 9.5 + \frac{(7,359)^2}{2(50,583)}$$
$$= 9.5 + 535.3$$
$$= 545 \text{ miles}$$

Clearly, the lunar ascent stage never reaches this height because the energy goes into orbital kinetic energy. At the 9.5-mile height, it must be traveling 7,359 miles per hour almost horizontally.

Substituting the value of H_{max} = 545 miles, as well as the other values, into the original equation results in

$$545 = \frac{(6,312)^2(\ln R)^2}{2(50,583)} - (6,312)(.121)\left(\frac{R}{R-1} \ln R - 1\right)$$

We merely have to solve for R to determine the payload ratio. The result is R = 7.2 for Moon gravity = .64 Earth gravity.

The equations for deriving the payload ratio assuming that the Moon has the same gravity as Earth are

$$H_{max} = H_{burnout} + \frac{U^2_{burnout}}{2G}$$

$$= 9.5 + \frac{(9,197)^2}{2(79,036)}$$

$$= 9.5 + 535.1$$

$$H_{max} = 545 \text{ miles}$$

and

$$545 = \frac{(6,312)^2(\ln R)^2}{2(79,036)} - (6,312)(.121)\left(\frac{R}{R-1}\ln R - 1\right)$$

Solving yields R = 18.2 for Moon gravity equal to Earth gravity. The same payload ratios can be used for the descent rockets.

APPENDIX F

DERIVATION OF THE RELATIVE HEIGHTS ATTAINED BY A JUMPING ASTRONAUT IN ONE-SIXTH GRAVITY CARRYING A BURDEN EQUAL TO HIS OWN WEIGHT AND THE SAME ASTRONAUT ON EARTH WITHOUT A BURDEN

From the conservation of energy it follows that

$$\tfrac{1}{2}MV^2 = MGH$$

where

M = the mass of a body
V = the initial velocity
G = the acceleration of gravity
H = the height attained

It follows that

$$H = \frac{V^2}{2G}$$

If G is the acceleration of gravity on Earth and V_e is the initial velocity on Earth, the height attained, H_e, is

$$H_e = \frac{V_e^2}{2G}$$

Under one-sixth gravity with initial velocity V_m, height H_m is

$$H_m = \frac{V_m^2}{2\left(\frac{G}{6}\right)} = \frac{6V_m^2}{2G}$$

The ratio of heights attained is

$$\frac{H_m}{H_e} = \frac{\frac{6V_m^2}{2G}}{\frac{V_e^2}{2G}}$$

$$\frac{H_m}{H_e} = \frac{6V_m^2}{V_e^2}$$

Therefore, if the initial velocities are the same, a body will rise 6 times as high as on Earth. If the initial velocity is twice as great, it will rise 24 times as high, and if 3 times as fast, it rises 54 times the Earth height.

Now, the ratio of initial velocities needs to be determined for an astronaut in one-sixth gravity with a burden equal to his own weight and the same astronaut on Earth without a burden. In each case, it is assumed that the jumping force is applied in an identical manner. This means that the distance through which the upward accelerating force acts is independent of the gravity effects. Therefore,

$$\tfrac{1}{2} a' t'^2 = \tfrac{1}{2} a t^2$$

where a' is the accelerating force and t' is the time it is applied under one-sixth gravity; and a and t are the corresponding values under normal gravity. Let F equal the force applied and W be the weight under normal gravity. Therefore, from the relation that acceleration equals force divided by mass,

$$a = \frac{F - W}{W}$$

In one-sixth gravity with a burden equal to the astronaut's weight,

$$a' = \frac{F - \frac{2W}{6}}{2W} = \frac{3F - W}{6W}$$

Appendix F

Since $a't'^2 = at^2$, then

$$t' = \left(\sqrt{\frac{a}{a'}}\right)t$$

Substituting the above values for a' and a yields

$$t' = \left(\sqrt{\frac{a}{a'}}\right)t = t\sqrt{\frac{\left(\frac{F-W}{W}\right)}{\left(\frac{3F-W}{6W}\right)}} = t\sqrt{\frac{6(F-W)}{3F-W}}$$

from which the initial velocity in one-sixth gravity equals

$$a't' = \left(\sqrt{\frac{6(F-W)}{3F-W}}\right)\left(\frac{3F-W}{6W}\right)t$$

Therefore,

$$(a't')^2 = \frac{(F-W)(3F-W)}{6W^2}t^2$$

Since

$$(at)^2 = \left(\frac{F-W}{W}\right)^2 t^2$$

it follows that the ratio of the square of the initial velocities is

$$\frac{(a't')^2}{(at)^2} = \frac{3F-W}{6(F-W)}$$

The ratio of the relative heights attained is

$$6\frac{V_m^2}{V_e^2} = 6\frac{(a't')^2}{(at)^2}$$

or,

$$6 \frac{3F - W}{6(F - W)} = \frac{3F - W}{F - W}$$

An astronaut weighing 185 pounds can jump vertically at least 18 inches with a force of 500 pounds on the Earth without a burden. Substituting these values in the above equation, yields

$$\frac{3(500) - 185}{500 - 185} = 4.17$$

This means that this same astronaut could jump over 4 times as high in one-sixth gravity carrying a burden equal to his own weight. This is over 6 feet off the ground.

APPENDIX G

DERIVATION OF THE MOON'S SURFACE GRAVITY USING JOHN YOUNG'S ESTIMATED VERTICAL JUMP OF 18 INCHES ON THE MOON ASSUMING HE CARRIED A BACKPACK LIFE-SUPPORT SYSTEM AND SPACESUIT EQUAL TO HIS OWN WEIGHT

The ratio of the relative heights attained assumes that John Young could have jumped 18 inches on the Earth without a burden, using the same effort which lifted him 18 inches on the Moon with the body weight burden. This ratio is $18/18 = 1.00$. As in Appendix F, it is equal to

$$\frac{\frac{V_m^2}{2G_m}}{\frac{V_e^2}{2G_e}} = \frac{G_e V_m^2}{G_m V_e^2}$$

Now, $\frac{V_m^2}{V_e^2}$ must be determined as a function of the ratio of Earth to Moon gravity, R. The ratio of the initial velocities squared is derived as a function of R using the same approach as in Appendix F.

$$a = \frac{F - W}{W}$$

$$a' = \frac{F - \frac{2W}{R}}{2W} \quad \text{where } R = \frac{G_e}{G_m}$$

$$t' = \left(\sqrt{\frac{a}{a'}}\right)t$$

$$t' = t\sqrt{\frac{2(F-W)}{\left(F-\frac{2W}{R}\right)}}$$

$$a't' = \left(\frac{F-\frac{2W}{R}}{2W}\right)\left(\sqrt{\frac{2(F-W)}{\left(F-\frac{2W}{R}\right)}}\right)t$$

$$(a't')^2 = \frac{(F-W)\left(F-\frac{2W}{R}\right)}{2W^2}$$

Since

$$\frac{(a't')^2}{(at)^2} = \frac{\left(F-\frac{2W}{R}\right)}{2(F-W)}$$

the height ratio equals

$$\frac{G_e\left(F-\frac{2W}{R}\right)}{2G_m(F-W)}$$

which equals

$$\frac{R\left(F-\frac{2W}{R}\right)}{2(F-W)}$$

Equating the expression to the actual height ratio results in

$$1.00 = \frac{R\left(500 - \frac{370}{R}\right)}{2(315)}$$

$$630 = 500R - 370$$

$$500R = 1000$$

$$R = \frac{1000}{500} = 2.00 = \frac{G_e}{G_m}$$

Therefore,

$$G_m = \frac{G_e}{2.00}$$

or,

$$G_m = .50 G_e$$

APPENDIX H

DERIVATION OF THE MOON'S SURFACE GRAVITY USING JOHN YOUNG'S ESTIMATED VERTICAL JUMP OF 18 INCHES ON THE MOON ASSUMING THAT HE CARRIED 75 POUNDS OF GEAR

The ratio of heights attained assumes that John Young could have jumped 18 inches on the Earth without a burden, using the same effort which lifted him 18 inches on the Moon with the burden. This ratio is $18/18 = 1.00$, and equals $\dfrac{G_e V_m^2}{G_m V_e^2}$. As in Appendix G,

$$a = \frac{F - W}{W}$$

$$a' = \frac{F - \dfrac{260}{R}}{260} \quad (260 = 185 + 75)$$

$$a = \frac{500 - 185}{185} = 1.7$$

$$t' = t\sqrt{\frac{1.7}{\dfrac{F - \dfrac{260}{R}}{260}}}$$

$$a't' = t\left(\frac{F - \frac{260}{R}}{260}\right)\bigg/\sqrt{\left(\frac{F - \frac{260}{R}}{260}\right)^{1.7}}$$

$$(a't')^2 = t^2\frac{\left(F - \frac{260}{R}\right)(1.7)}{260}$$

$$(at)^2 = (1.7)^2 t^2$$

$$\frac{(a't')^2}{(at)^2} = \frac{\left(F - \frac{260}{R}\right)}{260(1.7)}$$

$$(442)1.00 = R\left(F - \frac{260}{R}\right)$$

$$500R = 702$$

$$R = 1.404$$

$$G_m = .71 G_e$$

APPENDIX I

AN ANALYSIS OF ROVER PERFORMANCE IN ONE-SIXTH GRAVITY CONDITIONS

The coefficient of friction of rubber on a surface similar to pavement is approximately 0.6. This means that if the wheels of a vehicle were locked, it would require a horizontal force equal to 60 percent of the weight of the loaded vehicle to slide the wheels along the pavement. The Moon's surface is primarily loose dust and rock; therefore, the coefficient of friction for the Rover is assumed to be 0.5 to keep from understating the amount of friction actually encountered. The loaded Earth weight of 1,540 pounds would be only 257 pounds in one-sixth gravity. This would generate a total frictional force of 257 x 0.5 = 128 pounds, which is the horizontal force required to make the vehicle slide.

The maximum radius of curvature before the vehicle begins to slide is calculated by equating the centrifugal force to the frictional force. At the maximum Rover speed of 10.2 miles per hour or 15 feet per second, the equation is

$$\frac{MV^2}{R} = 128 \text{ pounds of force}$$

where

M = the vehicle mass
R = the maximum turning radius
V = the velocity

Since $M = \dfrac{W}{G} = \dfrac{1540}{32.2} = 47.8$ pounds of mass and $V =$

15 feet per second,

$$R = \frac{(47.8)(225)}{128} = 84 \text{ feet}$$

At 5 miles per hour,

$$R = \frac{(47.8)(53.773)}{128} = 20 \text{ feet}$$

The maximum braking force is 128 pounds, and inertial properties are independent of gravity conditions. From the relation Force equals Mass times Acceleration, or

$$F = MA, \quad A = \frac{F}{M}.$$

The Rover could only be decelerated at the rate of

$$A = \frac{128}{47.8} = 2.68 \text{ ft/sec}^2$$

From the relation Velocity equals Acceleration times Time, or $V = AT$, it would take

$$T = \frac{15}{2.676} = 5.6 \text{ seconds}$$

to bring the rover to a stop. The stopping distance would be $D = \frac{1}{2}AT^2$ or

$$D = \frac{1}{2}(2.676)(5.6)^2 = 42 \text{ feet}$$

FOOTNOTES

CHAPTER 1
1. Eugene M. Emme, ed., *The History of Rocket Technology,* (Detroit: Wayne State University Press, 1964), p. 86.
2. Ralph E. Lapp, *Man and Space—The Next Decade,* (New York: Harper & Brothers, 1961), p. 44.

CHAPTER 2
1. M. Vertregt, *Principles of Astronautics,* (New York: Elsevier Publishing Company, 1965), p. 135.
2. Franklyn M. Branley, *Exploration of the Moon,* (Garden City, New York: The Natural History Press, 1966), p. 53.
3. *U.S. on the Moon,* (Washington: U.S. News & World Report, 1969), p. 37.
4. Myrl H. Ahrendt, *The Mathematics of Space Exploration,* (New York: Holt, Rinehart and Winston, Inc., 1965), p. 55.
5. John A. Eisele, *Astrodynamics, Rockets, Satellites, and Space Travel,* (Washington: The National Book Company of America, 1967), p. 350.
6. *Collier's Encyclopedia,* 1961 ed., s.v. "Space Travel," p. 544.
7. *Encyclopaedia Britannica,* 14th ed., 1960, s.v. "Interplanetary Exploration," p. 530c.

CHAPTER 3
1. Martin Caidin, *The Moon: New World for Men,* (Indianapolis, Indiana: The Bobbs-Merrill Company, 1963), p. 111.
2. Ralph E. Lapp, *Man and Space—The Next Decade,* (New York: Harper & Brothers, 1961), p. 51.
3. Wernher von Braun and Frederick I. Ordway III, *History of Rocketry & Space Travel,* (New York: Thomas Y. Crowell Company, 1969), p. 191.
4. John Noble Wilford, *We Reach the Moon,* (New York: W.W. Norton & Company, Inc., 1969), p. 95.
5. "The Moon—A Giant Leap For Mankind," *Time,* July 25, 1969, p. 14.
6. Braun and Ordway, *History of Rocketry & Space Travel,* p. 238.
7. *Encyclopaedia Britannica,* 14th ed., 1973, s.v. "Space Exploration," p. 1045.
8. Wilford, *We Reach the Moon,* p. 54.
9. The Writers and Editors of the Associated Press with Manuscript by John Barbour, *Footprints on the Moon,* (The Associated Press, 1969), p. 201.

CHAPTER 4
1. Wernher von Braun, *Space Frontier,* (New York: Holt, Rinehart, and Winston, Inc., 1971), p. 215.
2. *Encyclopaedia Britannica,* 14th ed., 1973, s.v. "Space Exploration," p. 1045.

3. John Noble Wilford, *We Reach the Moon,* (New York: W.W. Norton & Company, Inc., 1969), p. 122.
4. Richard Lewis, *The Voyages of Apollo,* (New York: The New York Times Book Co., 1974), p. 104.

CHAPTER 5
1. James R. Berry, "How to Walk on the Moon," *Science Digest,* November 1967, p. 8.
2. *U.S. on the Moon,* (Washington: U.S. News & World Report, 1969), p. 54.
3. John Noble Wilford, *We Reach the Moon,* (New York: W.W. Norton & Company, Inc., 1969), pp. 298-305.
4. Richard Lewis, *The Voyages of Apollo,* (New York: The New York Times Book Co., 1974), p. 109.
5. Ibid., pp. 111-112.
6. "Intrepid on a Sun-drenched Sea of Storms," *Life,* December 12, 1969, p. 35.
7. Lloyd Mallan, *Suiting Up For Space,* (New York: The John Day Company, 1971), p. 239.
8. Alice J. Hall, "The Climb Up Cone Crater," *National Geographic,* July 1971, p. 148.
9. Lewis, *The Voyages of Apollo,* p. 187.
10. Ibid., p. 193.
11 Ibid., pp. 195-196.
12. Ibid., p.212.
13. Ibid., p.212.
14. Lawrence Maisak, *Survival on the Moon,* (New York: The Macmillan Company, 1966), pp. 133-134.
15. Lewis, *The Voyages of Apollo,* p. 248.
16. Ibid., p.257.
17. Ibid., pp. 259-260.
18. Ibid., p. 279.

CHAPTER 6
1. *U.S. on the Moon,* (Washington: U.S. News & World Report, 1969), pp. 51-55.
2. Franklyn M. Branley, *Exploration of the Moon,* (Garden City, New York: The Natural History Press, 1966), p. 34.
3. Wernher von Braun, *Space Frontier,* (New York: Holt, Rinehart and Winston, Inc., 1971), p. 156.

CHAPTER 7
1. Richard Lewis, *The Voyages of Apollo,* (New York: The New York Times Book Co., 1974), p. 67.
2. Ibid., p. 67.
3. Ibid., p. 107.
4. Ibid., p. 116.
5. Ibid., p. 116.

6. "Apollo 12 On The Moon," *Life,* December 12, 1969.
7. Paul M. Sears, "How Dead Is The Moon?," *Natural History,* February 1950, pp. 63-65.
8. V.A Firsoff, *Strange World of the Moon,* (New York: Basic Books, 1960), pp. 76-77.
9. Ibid., p. 81.
10. Ibid., p. 110
11. Charles Fort, *New Lands,* (New York: Ace Books, 1923), p. 42.
12. Firsoff, *Strange World of the Moon,* p. 129.
13. Lewis, *The Voyages of Apollo,* p. 134.
14. Howard Benedict, "Moon 'Eerie Sight', Apollo Chief Says," *The Indianapolis News,* July 19, 1969, p. 1.
15. Ibid., p. 1.

CHAPTER 8

1. David R. Scott, "What Is It Like to Walk on the Moon?," *National Geographic,* September 1973, p. 327.
2. Richard Lewis, *The Voyages of Apollo,* (New York: The New York Times Book Co., 1974), p. 218.
3. Ibid., p. 253.
4. William Gordon Allen, *Overlords, Olympians, and the UFO,* (Mokelumne Hill, California: Health Research, 1974), p. 110.
5. Lewis, *The Voyages of Apollo,* pp. 51-52.
6. Ibid., pp. 54-56.
7. "Glazing the Moon," *Time,* October 3, 1969, pp. 72-74.
8. V.A. Firsoff, *Strange World of the Moon,* (New York: Basic Books, 1960), p. 62.
9. Don Wilson, *Secrets of Our Spaceship Moon,* (New York: Dell Publishing Co., Inc., 1979), p. 33.
10. Joseph Goodavage, *Astrology: The Space Age Science,* (West Nyack, New York: Parker Publishing Co., Inc., 1966), p. 60.
11. Wilson, *Secrets of Our Spaceship Moon,* p. 107.
12. Joseph F. Goodavage, "What Strange and Frightening Discoveries Did Our Astronauts Make on the Moon?," *Saga,* March 1974, p. 36.

CHAPTER 9

1. "First Color Portrait of an Angry Earth," *Life,* November 10, 1967, p. 107.
2. Richard E. Byrd, "Our Navy Explores Antarctica," *National Geographic,* October 1947, pp. 429-522.
3. William Reed, *The Phantom of the Poles,* (Mokelumne Hill, California: Health Research, 1906), p. 50.
4. Fridtjof Nansen, *Farthest North,* (London: Harrison and Sons, 1897), Vol. I, p. 191.
5. Ibid., Vol. II, p. 138.
6. Ibid., Vol. I, p. 358.

7. Dr. Raymond Bernard, *The Hollow Earth*, (Secaucus, N.J.: Lyle Stuart Inc., 1969), pp. 28-29.
8. Ibid., p. 43.

CHAPTER 10
1. V.A. Firsoff, *Strange World of the Moon*, (New York: Basic Books, 1960), p. 80.
2. Joseph F. Goodavage, "Did Our Astronauts Find Evidence of UFOs on the Moon?," *Saga*, April 1974, p. 48.
3. George Leonard, *Somebody Else Is on the Moon*, (New York: Pocket Books, 1976), p. 61.
4. Charles Fort, *The Book of the Damned*, (New York: Ace Books, 1919), pp. 259-260.
5. Don Wilson, *Secrets of Our Spaceship Moon*, (New York: Dell Publishing Co., Inc., 1979), p. 207.
6. Timothy Green Beckley and Harold Salkin, "Apollo 12's Mysterious Encounter with Flying Saucers," *Saga UFO Special III*, 1972, p. 58.
7. Dr. Edward U. Condon, *Scientific Study of Unidentified Flying Objects*, (New York: Bantam Books, 1968), p. 194.
8. Don Wilson, *Our Mysterious Spaceship Moon*, (New York: Dell Publishing Co., Inc., 1975), p. 27.
9. Otto O. Binder, "Secret Messages From UFO's," *Saga UFO Special III*, 1972, p. 46.
10. Beckley and Salkin, "Apollo 12's Mysterious Encounter with Flying Saucers," *Saga UFO Special III*, p. 60.
11. Condon, *Scientific Study of Unidentified Flying Objects*, p. 207.
12. Martin Caidin, *Rendezvous in Space*, (New York: E.P. Dutton & Co., Inc., 1962), p. 124.
13. Binder, "Secret Messages From UFO's," *Saga UFO Special III*, p. 46.
14. Ibid., p. 46.
15. Ibid., p. 46.
16. John Noble Wilford, *We Reach the Moon*, (New York: W.W. Norton & Company, Inc., 1969), p. 219.
17. Wilson, *Secrets of Our Spaceship Moon*, p. 48.
18. Wilson, *Our Mysterious Spaceship Moon*, pp. 43-45.
19. Binder, "Secret Messages From UFO's," *Saga UFO Special III*, p. 46.
20. Ibid., p. 46.
21. Eric Faucher, Ellen Goodstein, and Henry Gris, "Alien UFOs Watched Our First Astronauts On Moon," *The National Enquirer*, September 11, 1979, p. 25.
22. Beckley and Salkin, "Apollo 12's Mysterious Encounter with Flying Saucers," *Saga UFO Special III*, p. 8.
23. Ibid., p. 58.
24. Ibid., p. 58.
25. Richard Lewis, *The Voyages of Apollo*, (New York: The New York Times Book Co., 1974), p. 252.
26. Wilson, *Secrets of Our Spaceship Moon*, p. 216.

CHAPTER 11

1. George Adamski, *Inside the Space Ships*, (New York: Abelard-Schuman, Inc., 1955), p. 76.
2. Ibid., p. 78.
3. Ibid., pp. 153-154.
4. Ibid., pp. 157-158.
5. Ibid., pp. 160-161.
6. Ibid., pp. 227-228.
7. Howard Menger, *From Outer Space to You,* (Clarksburg, W. Va.: Saucerian Books, 1959), p. 155.

CHAPTER 12

1. Arthur Fisher, "Science Newsfront - A Woolly Mammoth Story," *Popular Science,* April 1980, p. 15.

CHAPTER 13

1. Rho Sigma, *Ether-Technology: A Rational Approach to Gravity-Control,* (Lakemont, Georgia: Rho Sigma, 1977), p. 73.
2. Frank Edwards, *Strange World,* (New York: Bantam Books, 1964), p. 163.

CHAPTER 14

1. Howard Benedict, "Russians Press to Dominate Space Again," *The Oregonian,* September 30, 1979, p. A16.

BIBLIOGRAPHY

1. Adamski, George. *Inside the Space Ships.* New York: Abelard-Schuman, Inc., 1955.
2. Adamski, George; Leslie, Desmond. *Flying Saucers Have Landed.* New York: The British Book Centre, 1953.
3. Ahrendt, Myrl H. *The Mathematics of Space Exploration.* New York: Holt, Rinehart and Winston, Inc., 1965.
4. Allen, William Gordon. *Overlords, Olympians & the UFO.* Mokelumne Hill, California: Health Research, 1974.
5. *Analysis of Surveyor 3 Material and Photographs Returned by Apollo 12.* NASA SP-284. Washington: U.S. Government Printing Office, 1972.
6. Anderson, Commander William R., U.S.N.; Blair, Clay. *Nautilus 90 North.* Cleveland, Ohio: The World Publishing Company, 1959.
7. *Apollo 8: Man Around the Moon.* EP-66. Washington: U.S. Government Printing Office, 1968.
8. *Apollo 14: Preliminary Science Report.* NASA SP-272. Washington: U.S. Government Printing Office, 1971.
9. *Apollo 15: Preliminary Science Report.* NASA SP-289. Washington: U.S. Government Printing Office, 1972.
10. *Apollo 16: Preliminary Science Report.* NASA SP-315. Washington: U.S. Government Printing Office, 1972.
11. *Apollo 17: Preliminary Science Report.* NASA SP-330. Washington: U.S. Government Printing Office, 1973.
12. Arnold, Kenneth. "Fireflies and Flying Saucers," *Flying Saucers,* November 1962, pp. 29-31.
13. Barbour, John; the Writers and Editors of The Associated Press. *Footprints on the Moon.* The Associated Press, 1969.
14. Beckley, Timothy Green; Salkin, Harold. "Apollo 12's Mysterious Encounter with Flying Saucers," *Saga UFO Special III,* 1972, pp. 8-62.
15. Benedict, Howard. "Moon 'Eerie Sight' Apollo Chief Says," *The Indianapolis News,* July 19, 1969, p. 1.
16. _____. "Russians Press to Dominate Space Again," *The Oregonian,* September 30, 1979, p. A16.
17. Berlitz, Charles; Moore, William L. *The Roswell Incident.* New York: Grosset & Dunlap, 1980.
18. Bernard, Dr. Raymond. *The Hollow Earth.* Secaucus, N.J.: Lyle Stuart, Inc., 1969.
19. Berry, James R. "How to Walk on the Moon," *Science Digest,* November 1967, pp. 6-8.
20. Binder, Otto O. "Secret Messages From UFO's," *Saga UFO Special III,* 1972, pp. 24-48.
21. Boadella, David. *Wilhelm Reich - The Evolution of His Work.* Chicago: Henry Regnery Company, 1973.
22. Bodechtel, J.; Gierloff-Emden, H.G. *The Earth From Space.* New York: Arco Publishing Company, Inc., 1974.
23. Branley, Franklyn M. *Exploration of the Moon.* Garden City, New York: The Natural History Press, 1966.

24. Braun, Wernher von. *Space Frontier.* New York: Holt, Rinehart and Winston, Inc. 1971.
25. Braun, Wernher von; Ordway, Frederick I. III. *History of Rocketry & Space Travel.* New York: Thomas Y. Crowell Company, 1969.
26. Brown, Michael H. *PK, A Report on the Power of Psychokinesis, Mental Energy that Moves Matter.* Blauvelt, New York: Steinerbooks, 1976.
27. Brown, T.T. *Electrokinetic Apparatus.* Washington, D.C.: U.S. Patent Office, Patent #2,949,550, 1960.
28. _____. *Electrokinetic Generator.* Washington, D.C.: U.S. Patent Office, Patent #3,022,430, 1960.
29. _____. *Electrokinetic Apparatus.* Washington, D.C.: U.S. Patent Office, Patent #3,187,206, 1965.
30. Burr, Harold S. *The Fields of Life.* New York: Ballantine, 1973.
31. Byrd, Richard E. "Our Navy Explores Antarctica," *National Geographic,* October 1947, pp. 429-522.
32. _____. "All-Out Assault on Antarctica," *National Geographic,* August 1956, pp. 141-180.
33. Caidin, Martin. *Rendezvous in Space.* New York: E.P. Dutton & Co., Inc., 1962.
34. _____. *The Moon: New World For Men.* Indianapolis, Indiana: The Bobbs-Merrill Company, 1963.
35. *Collier's Encyclopedia,* 1961 ed., s.v. "Space Travel."
36. Collins, Michael. *Carrying the Fire.* New York: Farrar, Straus and Giroux, 1974.
37. Condon, Dr. Edward U. *Scientific Study of Unidentified Flying Objects.* New York: Bantam Books, 1968.
38. "Cosmonauts End Record Flight," *The Oregonian,* October 12, 1980, p. A8.
39. Crookes, Sir William. *Researches in the Phenomena of Spiritualism.* Litchfield, Connecticut: The Panthean Press, 1870.
40. Deyo, Stan. *The Cosmic Conspiracy.* Kalamunda, Western Australia: West Australian Texas Trading, 1978. (Available from Emissary Publications, P.O. Box 642, South Pasadena, Ca. 91030)
41. Dow, T.W. *Reshape Newton's Laws.* Washington: Celestial Press, 1965.
42. Driscoll, Robert B. *Unified Theory of Ether, Field and Matter.* Oakland, California: R.B. Driscoll, 1966.
43. Edwards, Frank. *Strange World.* New York: Bantam Books, 1964.
44. Eisele, John A. *Astrodynamics, Rockets, Satellites, and Space Travel.* Washington: The National Book Company of America, 1967.
45. Emme, Eugene M., ed. *The History of Rocket Technology.* Detroit: Wayne State University Press, 1964.
46. *Encyclopaedia Britannica,* 14th ed., 1960, s.v. "Interplanetary Exploration."
47. *Encyclopaedia Britannica,* 14th ed., 1973, s.v. "Space Exploration."
48. Faucher, Eric; Goodstein, Ellen; Gris, Henry. "Alien UFOs Watched Our First Astronauts on Moon," *The National Enquirer,* September 11, 1979, p. 25.
49. Fernandez, John P. *The Solution to the Riddle of Gravitation.* Hayward, California: John P. Fernandez, 1976. (Ordered from Fern's Science Books, Box 19010, Oakland, Ca. 94619)
50. Firsoff, V.A. *Strange World of the Moon.* New York: Basic Books, 1960.
51. "First Color Portrait of an Angry Earth," *Life,* November 10, 1967, p. 107.
52. Fisher, Arthur. "Science Newsfront - A Woolly Mammoth Story," *Popular Science,* April 1980, p. 15.

53. "Flight Plan of Apollo 11," *Time,* July 18, 1969, pp. 18-19.
54. Fort, Charles. *The Book of the Damned.* New York: Ace Books, 1919.
55. _____. *New Lands.* New York: Ace Books, 1923.
56. French, Bevan M. *The Moon Book.* New York: Penguin Books, 1977.
57. Gardner, Marshall B. *A Journey to the Earth's Interior.* Aurora, Illinois: Marshall B. Gardner, 1913.
58. "Glazing the Moon," *Time,* October 3, 1969, pp. 72-74.
59. Goodavage, Joseph F. *Astrology: The Space Age Science.* West Nyack, New York: Parker Publishing Co., 1966.
60. _____. "What Strange and Frightening Discoveries Did Our Astronauts Make on the Moon?," *Saga,* March 1974, pp. 30-39.
61. _____. "Did Our Astronauts Find Evidence of UFOs on the Moon?," *Saga,* April 1974, pp. 30-51.
62. Hall, Alice J. "Apollo 14: The Climb Up Cone Crater," *National Geographic,* July 1971, pp. 136-148.
63. Hall, R. Cargill. *Lunar Impact - A History of Project Ranger.* Washington: U.S. Government Printing Office, 1977.
64. Hamilton, William F. *Center of the Vortex.* Los Angeles, California: Nexus & Nexus News, 1979.
65. "Intrepid on a Sun-drenched Sea of Storms," *Life,* December 12, 1969, pp. 34-39.
66. Jordan, Pascual. *The Expanding Earth.* New York: Pergamon Press Inc., 1971.
67. Kaysing, Bill; Reid, Randy. *We Never Went to the Moon.* Fountain Valley, California: Eden Press, 1976.
68. Kemp, Jack; Aspin, Les. *How Much Defense Spending Is Enough?.* Washington: American Enterprise Institute for Public Policy Research, 1976.
69. Kervran, Louis. *Biological Transmutations.* Brooklyn, New York: Swan House Publishing Co., 1972.
70. Kilner, Walter J. *The Human Aura.* New York: University Books, 1965.
71. Kor, Peter. "Mystery Clouds Over the Poles," *Flying Saucers,* September 1963, pp. 59-61.
72. _____. "The Inner Earth Theory," *Flying Saucers,* November 1962, pp. 40-44.
73. Krippner, Stanley; Rubin, Daniel. *The Kirlian Aura.* Garden City, New York: Anchor Press/Doubleday, 1974.
74. Lapp, Ralph E. *Man and Space—The Next Decade.* New York: Harper & Brothers, 1961.
75. Lear, John. "The Hidden Perils of a Lunar Landing," *Saturday Review,* June 7, 1969, pp. 47-54.
76. Leonard, George. *Somebody Else Is on the Moon.* New York: Pocket Books, 1976.
77. Lewis, Richard. *The Voyages of Apollo.* New York: The New York Times Book Co., 1974.
78. Lloyd, John Uri. *Etidorhpa or The End of Earth.* Cincinnati, Ohio: The Robert Clarke Company, 1896.
79. "Lunar Blastoff," *The Oregonian,* April 24, 1972, p. 1.
80. "Lunar Salute to Flag," *The Oregonian,* April 22, 1972, p. 1
81. Maisak, Lawrence. *Survival on the Moon.* New York: The Macmillan Company, 1966.
82. Mallan, Lloyd. *Suiting Up For Space.* New York: The John Day Company, 1971.

83. Mann, W. Edward. *Orgone, Reich & Eros.* New York: Simon & Schuster, 1973.
84. Masursky, H.; Colton, G.W.; El-Baz, Farouk; eds. *Apollo Over the Moon - A View From Orbit.* NASA SP-362. Washington: U.S. Government Printing Office, 1978.
85. McCauley, John F. *Moon Probes.* Boston: Little, Brown and Company, 1969.
86. McGuire, Martin C. *Secrecy and the Arms Race.* Cambridge, Massachusetts: Harvard University Press, 1965.
87. Menger, Howard. *From Outer Space to You.* Clarksburg, West Virginia: Saucerian Books, 1959.
88. "The Moon—A Giant Leap For Mankind," *Time,* July 25, 1969, pp. 10-14.
89. "More Photos of the North Pole," *Flying Saucers,* December 1970, pp. 16-20.
90. "Mystery Force Causes Satellites to Lose Weight," *Flying Saucers,* January 1962, pp. 21-28.
91. Nansen, Fridtjof. *Farthest North.* London: Harrison and Sons, 1897.
92. Pearsall, Ronald. *The Table-Rappers.* London: Michael Joseph Ltd., 1972.
93. Pike, Richard J. *Geometric Interpretation of Lunar Craters.* Geological Survey Professional Paper 1046-C. Washington: U.S. Government Printing Office, 1980.
94. Puharich, Andrija, ed. *The Iceland Papers - Selected Papers on Experimental and Theoretical Research on the Physics of Consciousness.* Amherst, Wisconsin: Essentia Research Associates, 1979.
95. Reed, William. *The Phantom of the Poles.* Mokelumne Hill, California: Health Research, 1906.
96. Reich, Wilhelm. *The Cancer Biopathy.* New York: Farrar, Straus and Giroux, 1948.
97. _____. *Selected Writings.* New York: Farrar, Straus and Giroux, 1951.
98. Reichenbach, Baron Karl von. *Researches on Magnetism, Electricity, Heat, Light, Crystallization, and Chemical Attraction, in Their Relations To The Vital Force.* Secaucus, New Jersey: University Books, 1850.
99. _____. *Reichenbach's Letters on Od and Magnetism.* Mokelumne Hill, California: Health Research, 1852.
100. Scott, David R. "What Is It Like to Walk on the Moon?," *National Geographic,* September 1973, pp. 326-331.
101. Sears, Paul M. "How Dead Is the Moon?," *Natural History,* February 1950, pp. 62-65.
102. Shils, Edward A. *The Torment of Secrecy.* New York: Arcturus Books, 1956.
103. Sigma, Rho. *Ether-Technology: A Rational Approach to Gravity-Control.* Lakemont, Georgia: Rho Sigma, 1977. (Ordered from the Association for Research and Enlightenment, Inc., P.O. Box 595, Virginia Beach, Va. 23451)
104. Simmons, Gene. *On the Moon With Apollo 16.* EP-95. Washington: U.S. Government Printing Office, April 1972.
105. _____. *On the Moon With Apollo 17.* EP-101. Washington: U.S. Government Printing Office, December 1972.
106. Sitchin, Zecharia. *The 12th Planet.* New York: Avon Books, 1976.
107. Sokolow, Leonid. *A Dual Ether Universe.* Hicksville, New York: Exposition Press, 1977.
108. Sprague, Gale C. "Visual Studies of the Aurora," *Sky and Telescope,* June 1968, pp. 346-349.
109. Swenson, Loyd S., Jr. *The Ethereal Aether.* Austin, Texas: University of Texas Press, 1972.
110. Talbott, Stephen A., ed. *Velikovsky Reconsidered.* New York: Warner Books, 1966.

Bibliography

111. *U.S. on the Moon*. Washington: U.S. News & World Report, 1969.
112. Velikovsky, Immanuel. *Worlds In Collision*. New York: Doubleday & Co., Inc., 1950.
113. _____. *Earth in Upheaval*. New York: Dell Publishing Co., Inc., 1955.
114. Vertregt, M. *Principles of Astronautics*. New York: Elsevier Publishing Company, 1965.
115. Very, Frank W. *The Luminiferous Ether*. Boston: The Four Seas Company, 1919.
116. "Where the Reader Has His Say," (Polar Entrance Editorial), *Flying Saucers*, September 1970, pp. 24-40.
117. "Where the Reader Has His Say," (Polar Entrance Editorial), *Flying Saucers*, December 1970, pp. 29-40.
118. Wilford, John Noble. *We Reach the Moon*. New York: W.W. Norton & Company, Inc., 1969.
119. Wilson, Don. *Our Mysterious Spaceship Moon*. New York: Dell Publishing Co., Inc., 1975.
120. _____. *Secrets of Our Spaceship Moon*. New York: Dell Publishing Co., Inc., 1979.